College Algebra Formula Sheet and Key Points

Quick Study Guide and Test Prep Book for Beginners and Advanced Students + Two Practice Tests

Dr. Abolfazl Nazari

PUBLISHED BY EFFORTLESS MATH EDUCATION

EFFORTLESSMATH.COM

College Algebra Formula Sheet and Key Points

2024

WELCOME to College Algebra Formula Sheet and Key Points. Aimed at demystifying College Algebra, this guide is your key to understanding core concepts and grasping the essential formulas required. By distilling complex mathematics into concise, straightforward points, we make learning accessible and effective. Whether you're preparing the night before or seeking a last-minute review, this book is crafted to enhance recall and simplify your study process.

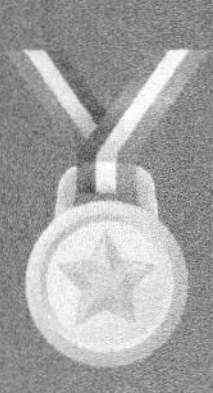

Embark on this educational journey with us and discover a smoother path to mastering College Algebra.

College Algebra Formula Sheet and Key Points is for students that are eager to learn College Algebra essentials in the fastest way. It distills complex topics down to bite-sized, manageable key points, alongside must-know formulas, facilitating quick learning and retention.

Experience a groundbreaking approach to math learning, designed for quick comprehension and lasting memory of crucial concepts. Understanding the challenge of locating all necessary formulas in one spot, we've dedicated a chapter to compiling every formula vital for the College Algebra examination.

What is included in this book

- ☑ Additional online resources for extended practice and support.
- ☑ A short manual on how to use this book.
- ☑ Coverage of all College Algebra subjects and topics tested.
- ☑ A dedicated chapter listing all the necessary formulas.
- ☑ Two comprehensive, full-length practice test with thorough answer explanations.

Effortless Math's College Algebra Online Center

Effortless Math Online College Algebra Center offers a complete study program, including the following:

- ☑ *Step-by-step instructions on how to prepare for the College Algebra test*
- ☑ *Numerous College Algebra worksheets to help you measure your math skills*
- ☑ *Complete list of College Algebra formulas*
- ☑ *Video lessons for all College Algebra topics*
- ☑ *Full-length College Algebra practice tests*

Visit `EffortlessMath.com/CollegeAlgebra` to find your online College Algebra resources.

Scan this QR code

(No Registration Required)

Tips for Making the Most of This Book

This book is your fastest route to mastering College Algebra. It distills the subject down to key points and includes the formulas you need to remember for the College Algebra topics. Here's what makes it special:

- Concise content that focuses on the essentials.
- A dedicated chapter of formulas to remember, covering all topics.
- Two practice tests at the end of the book to assess your knowledge and readiness.

Mathematics can be approachable and easy with the right tools and mindset. Our goal is to simplify math for you, focusing on what's necessary. Using the key points and the formula sheets, you can quickly review and remember the most important information.

Here's how to leverage this book effectively:

Understand and Remember Key Points

Each math topic is built around core ideas or concepts. We've highlighted key points in every topic as mini-summaries of critical information. Don't skip these!

Formulas to Remember

At the end of each topic, we've included a list of formulas to remember. These are the essential formulas you need to know for College Algebra. Make sure to review these regularly.

In summary:

- ***Key Points***: Essential summaries of major concepts.
- ***Formulas to Remember***: Each topic ends with a list of formulas to remember.

Once you have finished all chapters, review the formula sheet before taking the practice tests. This will help you recall and apply the formulas effectively.

The Practice Tests

The practice tests at the end of the book are a great way to gauge your readiness and identify areas needing more attention. Time yourself and simulate the actual test environment as closely as possible. After completing the practice tests, review your answers to find areas that require additional practice.

Effective Test Preparation

A solid test preparation plan is key. Beyond understanding concepts, strategic study and practice under exam conditions are vital.

- **Begin Early**: Start studying well in advance.
- **Daily Study Sessions**: Regular, short study periods enhance retention.
- **Active Note-Taking**: Helps internalize concepts and improves focus.
- **Review Challenges**: Focus extra time on difficult topics.
- **Practice**: Use end-of-chapter problems and additional resources for extensive practice.

Pick the Right Study Environment and Materials

In addition to this book, consider using other resources for College Algebra. Here are some suggestions: if you need more explanations of the topics, you can use the *College Algebra Made Easy* study guide. For additional practice, try our *College Algebra Workbook*.

Contents

1. Fundamentals and Building Blocks

1.1 Order of Operations

Key Point

PEMDAS, the acronym for the order of operations, stands for Parentheses, Exponents, followed by Multiplication and Division (left to right), and then Addition and Subtraction (left to right).

Formula To Remember

1 *Order of Operations (PEMDAS)* Parentheses → Exponents → Multiplication/Division (left to right) → Addition/Subtraction (left to right).

1.2 Scientific Notation

Key Point

Scientific notation is a method used to represent very large or very small numbers in a simplified manner. It is written as $m \times 10^n$ where m is a number greater than 1 and less than 10, and n is an exponent.

Key Point

To convert a number from scientific notation to standard form, move the decimal point n places to the right if n is positive, moving it to the left if n is negative.

Formula To Remember

1. *Writing in Scientific Notation* ☞ $m \times 10^n$ where $1 \le m < 10$ and n is an integer.

1.3 Rules of Exponents

Key Point

An exponent is simply a shorthand way for repeated multiplication of the same number by itself. It is written as x^n, where x is the base and n is the exponent.

Key Point

Exponent rules simplified:

Rule 1, Product Rule: $x^m \times x^n = x^{(m+n)}$.

Rule 2, Quotient Rule: $\frac{x^m}{x^n} = x^{(m-n)}$.

Rule 3, Power of Power: $(x^m)^n = x^{mn}$.

Rule 4, Product to Power: $(xy)^n = x^n y^n$.

Rule 5, Quotient to Power: $\left(\frac{x}{y}\right)^n = \frac{x^n}{y^n}$.

Formula To Remember

1. *Product of Powers* ☞ $x^m \times x^n = x^{(m+n)}$
2. *Quotient of Powers* ☞ $\frac{x^m}{x^n} = x^{(m-n)}$
3. *Power of a Power* ☞ $(x^m)^n = x^{mn}$
4. *Power of a Product* ☞ $(xy)^n = x^n y^n$
5. *Power of a Quotient* ☞ $\left(\frac{x}{y}\right)^n = \frac{x^n}{y^n}$

1.4 Evaluating Expressions

Key Point

To evaluate an algebraic expression, substitute a number for each variable and perform the arithmetic operations.

1.5 Simplifying Algebraic Expressions

Key Point

Like terms are those with both the same variable and the same exponent. Combining these terms allows for the simplification of an algebraic expression.

1.6 Sets

Key Point

Each object in a set is unique. If an object appears more than once in a set, it is still considered as a single entry. The order of elements in a set does not matter.

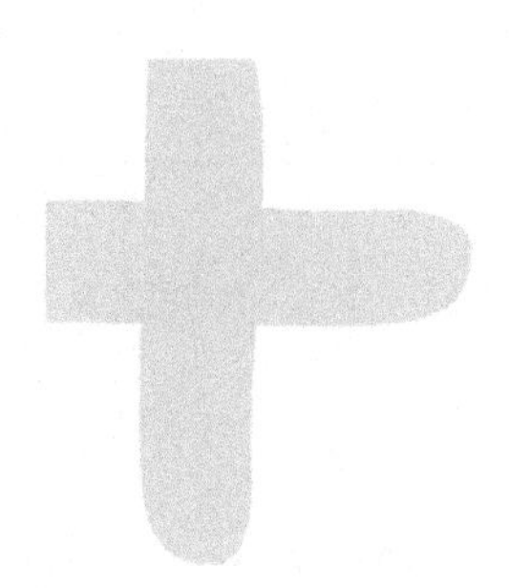

2. Equations and Inequalities

2.1 Solving Multi–Step Equations

Key Point

In multi-step equations, it is essential to maintain balance by applying identical operations to both sides. This principle forms the cornerstone of equation solving.

2.2 Slope and Intercepts

Key Point

The slope m of a line through points $A(x_1, y_1)$ and $B(x_2, y_2)$ is the ratio of vertical to horizontal change: $m = \frac{y_2 - y_1}{x_2 - x_1}$.

Key Point

A linear function is represented by $y = mx + b$, where m is the slope and b is the y-intercept. Such a linear equation is called slope-intercept form.

Formula To Remember

1. *Slope of a Line Through Two Points* $m = \frac{y_2 - y_1}{x_2 - x_1}$ where $A(x_1, y_1)$ and $B(x_2, y_2)$ are two distinct points on the line.
2. *Slope-Intercept Form of a Line* $y = mx + b$ where m is the slope and $(0, b)$ is the y-intercept.

2.3 Using Intercepts

Key Point

The x-intercept, where $y = 0$, is the x value where the line crosses the x-axis, and the y-intercept, where $x = 0$, is the y value where the line crosses the y-axis.

Formula To Remember

1. *x-Intercept* Set $y = 0$ in $ax + by = c$. Solve for x to get x-intercept $(x, 0)$.
2. *y-Intercept* Set $x = 0$ in $ax + by = c$. Solve for y to get y-intercept $(0, y)$.

2.4 Transforming Linear Functions

Key Point

Three basic transformations for linear functions include translation (shifting the line up or down), rotation (altering the steepness), and reflection (reflecting across the y-axis).

Key Point

Linear function $y = mx + b$ can undergo three transformations:

1) Translation: Changing b shifts the graph vertically.

2) Rotation: Modifying m alters steepness and causes rotation around $(0, b)$.

3) Reflection: Negating m reflects the function across y-axis.

These do not change the graph's shape but affect position, size, and steepness.

Formula To Remember

1. *Translation: Vertical Shift of a Line* $y = mx + (b \pm k)$ translates the line k units vertically.
2. *Rotation: Changing the Slope* $y = (m \pm k)x + b$ rotates the line around the point $(0, b)$, altering steepness.
3. *Reflection: Flipping Across the y-axis* $y = -mx + b$ reflects the line across the y-axis.

2.5 Solving Inequalities

Key Point

When you multiply or divide both sides by a negative number while solving inequalities, remember to flip the inequality sign.

2.6 Graphing Linear Inequalities

Key Point

For graphing linear inequalities, use a solid line when the inequality includes equality ($\leq$ or $\geq$), and a dotted line for strict inequalities ($<$ or $>$).

2.7 Solving Compound Inequalities

Key Point

When solving compound inequalities joined by "and", find the intersection of the individual solutions.

Key Point

When solving compound inequalities, joined by "or", form the union of the individual solutions.

Formula To Remember

1. *Conjunction (AND)* Solve A and B separately, then find the intersection. For $A \cap B$, x must satisfy both A and B.
2. *Disjunction (OR)* Solve A and B separately, then find the union. For $A \cup B$, x satisfies either A or B.
3. *Graphing Compound Inequalities* Use open circles for exclusive ($<$, $>$) and closed circles for inclusive ($\leq$, $\geq$). Draw lines to show the range of x values that satisfy the inequality.

2.8 Solving Absolute Value Equations

Key Point

Solving absolute value equations involves creating two related equations without absolute value signs.

Key Point

Absolute value equations like $|ax+b|=c$ (with constants $a \neq 0$, b, and $c \geq 0$) lead to two scenarios: $ax+b=c$ and $ax+b=-c$. Solve both for x and verify the solutions.

1. *Standard Absolute Value Equation* ☞ $|ax+b|=c$, then we have $x=\frac{c-b}{a}$ and $x=\frac{-c-b}{a}$

2.9 Solving Absolute Value Inequalities

Key Point

When working with absolute value inequalities, the inequality's direction dictates the separation of the absolute value expression into different inequalities. For $|ax+b| \leq c$, the solution falls within the interval $[-c,c]$, whereas for $|ax+b| \geq c$, the solution lies outside the interval $[-c,c]$.

Formula To Remember

1. *Solving* "$|ax+b| \leq c$" ☞ $-c \leq ax+b \leq c$
2. *Solving* "$|ax+b| \geq c$" ☞ $ax+b \geq c$ or $ax+b \leq -c$

2.10 Graphing Absolute Value Inequalities

Key Point

The inequality's direction ($\leq$, $\geq$, $<$, $>$) recorresponds to whether a circle on the number line plot should be open (exclusivity) or closed (inclusivity).

2.11 Solving Systems of Equations

Key Point

The elimination method employs the addition property of equality to systematically eliminate one variable from a system of equations, simplifying the solving process.

2.12 Solving Special Systems

Key Point

Parallel lines indicate a system of equations with no solution while overlapping lines indicate a system of equations with an infinite number of solutions.

Formula To Remember

1. *Condition for Parallel Lines* Two equations $y = m_1x + b_1$ and $y = m_2x + b_2$ are parallel (and have no solution) if $m_1 = m_2$ and $b_1 \neq b_2$.
2. *Condition for Overlapping Lines* Two equations $y = m_1x + b_1$ and $y = m_2x + b_2$ overlap (and have an infinite number of solutions) if $m_1 = m_2$ and $b_1 = b_2$.

2.13 Systems of Equations Word Problems

Key Point

The solution to a system of equations in word problems corresponds to the solution of the problem in the real-world context.

3. Quadratic Function

3.1 Solving a Quadratic Equation

Key Point

In a quadratic equation, $ax^2+bx+c=0$, the solutions are given by $x=\frac{-b\pm\sqrt{b^2-4ac}}{2a}$, where b^2-4ac is known as the discriminant.

Key Point

The discriminant determines the nature of the solutions as follows:

1) $b^2-4ac>0$: Two distinct real roots.

2) $b^2-4ac=0$: Exactly one real root (repeated).

3) $b^2-4ac<0$: No real roots.

Formula To Remember

1. *Standard Form of a Quadratic Equation* $ax^2+bx+c=0$ where $a\neq 0$.
2. *Discriminant* $D=b^2-4ac$
3. *Quadratic Formula* $x=\frac{-b\pm\sqrt{D}}{2a}$ where $a\neq 0$.
4. *Nature of Roots* If $D>0$, two distinct real roots; if $D=0$, one real root; if $D<0$, no real roots.

3.2 Graphing Quadratic Functions

Key Point

The standard form of a quadratic function is $y = ax^2 + bx + c$, where the vertex of the function is located at $x = -\frac{b}{2a}$.

Key Point

The vertex form of a quadratic function is $y = a(x-h)^2 + k$, where (h,k) represents the vertex of the function. The axis of symmetry is $x = h$.

Formula To Remember

1. *Standard Form of a Quadratic Function* ☞ $y = ax^2 + bx + c$ where $a \neq 0$, b, and c are constants.
2. *Vertex Form of a Quadratic Function* ☞ $y = a(x-h)^2 + k$ $(a \neq 0)$ where (h,k) is the vertex and $x = h$ is the axis of symmetry.
3. *Vertex from Standard Form* ☞ Vertex (h,k) is found by $h = -\frac{b}{2a}$ and k by substituting $x = h$ into the standard form.

3.3 Axis of Symmetry of Quadratic Functions

Key Point

In a quadratic function given in the standard form, $y = ax^2 + bx + c$, the axis of symmetry is found using the formula $x = -\frac{b}{2a}$.

Key Point

In a quadratic function given in vertex form, $y = a(x-h)^2 + k$, the axis of symmetry is $x = h$.

Formula To Remember

1. *Axis of Symmetry (Standard Form)* ☞ $x = -\frac{b}{2a}$ for the quadratic function $y = ax^2 + bx + c$.
2. *Axis of Symmetry (Vertex Form)* ☞ $x = h$ for the quadratic function $y = a(x-h)^2 + k$.

3.4 Solve a Quadratic Equation by Graphing

Key Point

To solve a quadratic equation by graphing, create two new equations from the original, graph them, and find the intersection points, which represent the solutions.

3.5 Solving Quadratic Equations with Square Roots

Key Point

To solve $ax^2 = c$, isolate the square term, take the square root on both sides, and include a "$\pm$" sign, representing both the positive and negative roots.

Key Point

To solve $(ax+b)^2 = c$, take the square root on both sides, leading to $ax+b = \pm\sqrt{c}$. Solve for x from this equation.

Formula To Remember

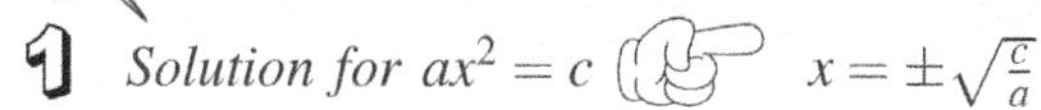

1. *Solution for* $ax^2 = c$ ☞ $x = \pm\sqrt{\frac{c}{a}}$
2. *Solution for* $(ax+b)^2 = c$ ☞ $x = \frac{-b\pm\sqrt{c}}{a}$

3.6 Build Quadratics from Roots

Key Point

To create a quadratic equation from its roots, use the reverse factoring method. The standard equation is $x^2 - (\alpha+\beta)x + \alpha\beta = 0$, where α and β are the roots, and it satisfies the quadratic form with the sum of roots equaling the coefficient of x and the product of roots as the constant term.

Formula To Remember

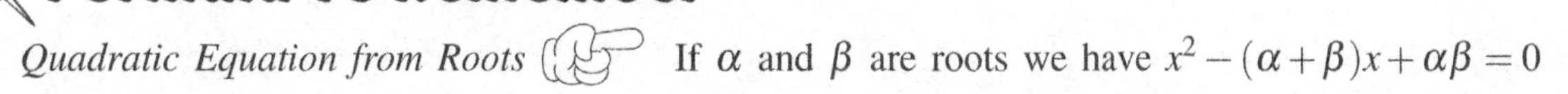

1. *Quadratic Equation from Roots* ☞ If α and β are roots we have $x^2 - (\alpha+\beta)x + \alpha\beta = 0$

3.7 Solving Quadratic Inequalities

Key Point

In solving a quadratic inequality, begin by converting it into an equation and finding its roots. The solutions found will divide the number line into intervals.

3.8 Graphing Quadratic Inequalities

Key Point

To graph a quadratic inequality, start by plotting the corresponding quadratic equation's graph, the parabola $y = ax^2 + bx + c$. The roots of this equation represent the points where the graph intersects the x-axis, dividing it into intervals. Test a point in each interval.

3.9 Factoring the Difference of Two Perfect Squares

Key Point

The formula for factoring the difference of two perfect squares is

$$a^2 - b^2 = (a-b)(a+b).$$

To factor expressions involving perfect squares, find the square roots of the perfect squares in the expression and use this formula.

Formula To Remember

1 *Difference of Two Perfect Squares* 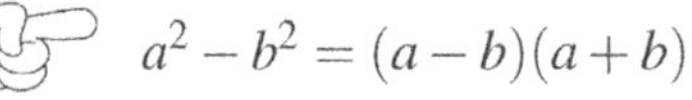 $a^2 - b^2 = (a-b)(a+b)$

4. Complex Numbers

4.1 Adding and Subtracting Complex Numbers

Key Point

The rule for adding and subtracting complex numbers is given as follows:

Addition: $(a+bi)+(c+di)=(a+c)+(b+d)i$

Subtraction: $(a+bi)-(c+di)=(a-c)+(b-d)i$

Formula To Remember

1. *Addition of Complex Numbers* $(a+bi)+(c+di)=(a+c)+(b+d)i$
2. *Subtraction of Complex Numbers* $(a+bi)-(c+di)=(a-c)+(b-d)i$

4.2 Multiplying and Dividing Complex Numbers

Key Point

The rule for multiplying complex numbers $(a+bi)$ and $(c+di)$ is:

$$(a+bi)(c+di)=(ac-bd)+(ad+bc)i.$$

Key Point

The conjugate of a complex number $(a+bi)$ is $(a-bi)$, which only changes the sign of the imaginary part.

Key Point

The process for dividing complex numbers $\frac{a+bi}{c+di}$ is:

$$\frac{a+bi}{c+di} \times \frac{c-di}{c-di} = \frac{ac+bd}{c^2+d^2} + \frac{bc-ad}{c^2+d^2}i.$$

Formula To Remember

1. *Multiplying Complex Numbers* ☞ $(a+bi)(c+di) = (ac-bd)+(ad+bc)i$
2. *Conjugate of a Complex Number* ☞ For $z = a+bi$ the conjugate, $\bar{z}$, is $a-bi$
3. *Dividing Complex Numbers* ☞ $\frac{a+bi}{c+di} = \frac{a+bi}{c+di} \times \frac{c-di}{c-di} = \frac{ac+bd}{c^2+d^2} + \frac{bc-ad}{c^2+d^2}i$

4.3 Rationalizing Imaginary Denominators

Key Point

To rationalize the denominator of a complex fraction $\frac{a+bi}{c+di}$, multiply both numerator and denominator by the conjugate of the denominator, $(c-di)$.

Formula To Remember

1. *Rationalization of a Complex Fraction* ☞ Multiply by the conjugate: $\frac{a+bi}{c+di} \times \frac{c-di}{c-di}$

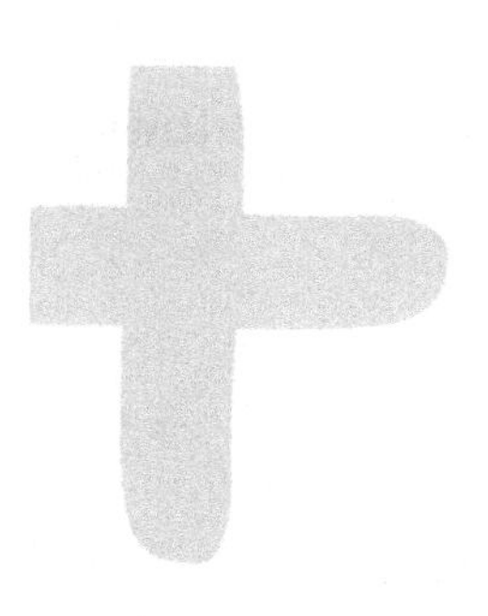

5. Matrices

5.1 Using Matrices to Represent Data

Key Point

Matrix representation is a method to store, sort, and operate data, especially when dealing with multivariate systems.

Key Point

Matrices come in various sizes, and their dimensions are described as $m \times n$, where m is the number of rows and n is the number of columns.

5.2 Adding and Subtracting Matrices

Key Point

In matrix addition and subtraction, matrices must have the same dimensions, and the operation is performed element-wise.

Formula To Remember

1 *Matrix Addition and Subtraction* Given matrices A and B of the same dimensions $m \times n$, the sum or difference C is defined as: $C = A \pm B$ where $C_{ij} = A_{ij} \pm B_{ij}$ for all $1 \leq i \leq m$ and $1 \leq j \leq n$.

5.3 Matrix Multiplication

Key Point

Matrix multiplication requires the first matrix to have the same number of columns as the second matrix has rows.

Key Point

In matrix multiplication, each row of the first matrix is multiplied with each column of the second matrix, and the resulting products are summed.

Formula To Remember

1. *Matrix Multiplication Criteria* The number of columns in the first matrix must be equal to the number of rows in the second matrix.

2. *Matrix Multiplication Process* If A is an $m \times n$ matrix and B is an $n \times p$ matrix, then $C = A \times B$ is an $m \times p$ matrix with entries $c_{ij} = a_{i1}b_{1j} + a_{i2}b_{2j} + \cdots + a_{in}b_{nj}$.

5.4 Finding Determinants of a Matrix

Key Point

The determinant is a fundamental property of matrices and is defined only for square matrices.

Key Point

For 2×2 matrices like $A = \begin{bmatrix} a & b \\ c & d \end{bmatrix}$, the determinant is $ad - bc$.

Key Point

For 3×3 matrices like $A = \begin{bmatrix} a & b & c \\ d & e & f \\ g & h & i \end{bmatrix}$, the determinant is calculated as

$$|A| = a(ei - fh) - b(di - fg) + c(dh - eg).$$

Formula To Remember

1 *Determinant of a 2 × 2 Matrix* For $A = \begin{bmatrix} a & b \\ c & d \end{bmatrix}$, $|A| = ad - bc$.

2 *Determinant of a 3 × 3 Matrix* For $A = \begin{bmatrix} a & b & c \\ d & e & f \\ g & h & i \end{bmatrix}$, $|A| = a(ei - fh) - b(di - fg) + c(dh - eg)$.

5.5 The Inverse of a Matrix

Key Point

A matrix A has an inverse, denoted as A^{-1}, if and only if it is square (having the equal number of rows and columns) and non-singular (with a non-zero determinant).

Key Point

The inverse of a 2 × 2 matrix $A = \begin{bmatrix} a & b \\ c & d \end{bmatrix}$ is $A^{-1} = \frac{1}{ad-bc} \begin{bmatrix} d & -b \\ -c & a \end{bmatrix}$.

Formula To Remember

1 *Condition for the Existence of an Inverse* A matrix A has an inverse A^{-1} if it is square and non-singular (i.e., $\det(A) \neq 0$).

2 *Inverse of a 2 × 2 Matrix* For $A = \begin{bmatrix} a & b \\ c & d \end{bmatrix}$, its inverse $A^{-1} = \frac{1}{ad-bc} \begin{bmatrix} d & -b \\ -c & a \end{bmatrix}$.

5.6 Solving Linear Systems with Matrix Equations

Key Point

A system of linear equations can be represented as a matrix equation of the form $AX = B$, facilitating easier solutions with matrix operations.

Key Point

The inverse A^{-1} is crucial for solving the matrix equation. To find X, multiply A^{-1} with B.

Formula To Remember

1. *Writing System of Equations as Matrix Equation* $AX = B$ where A is the coefficient matrix, X is the variable matrix, and B is the constant matrix.
2. *Matrix Equation Solution Using Inverse* $X = A^{-1}B$, if $|A| \neq 0$.

6. Polynomial Operations

6.1 Writing Polynomials in Standard Form

Key Point

Polynomial $f(x) = a_nx^n + \cdots + a_1x + a_0$ with $n \geq 0$ and coefficients $a_0, \cdots, a_n$ is ordered by decreasing degree.

Key Point

Rewriting a polynomial in standard form clarifies its structure, degree, and simplifies arithmetic operations and solving.

Formula To Remember

1. *Standard Form of a Polynomial* $f(x) = a_nx^n + a_{n-1}x^{n-1} + \cdots + a_1x + a_0$ where each a_i is a coefficient and n is the degree.
2. *Degree of the Polynomial* The degree is the highest exponent of x.

6.2 Simplifying Polynomials

Key Point

Simplification of polynomials through the combination of like terms, which are terms with matching variables and exponents, lessens their complexity.

Formula To Remember

1 *Combining Like Terms* $ax^ny^m + bx^ny^m = (a+b)x^ny^m$

6.3 Adding and Subtracting Polynomials

Key Point

For polynomials, addition combines like terms, and subtraction also combines like terms after adjusting for sign changes.

Formula To Remember

1 *Adding and Subtracting Polynomials* $(a_nx^n + a_{n-1}x^{n-1} + \ldots + a_1x + a_0) \pm (b_nx^n + b_{n-1}x^{n-1} + \ldots + b_1x + b_0) = (a_n \pm b_n)x^n + (a_{n-1} \pm b_{n-1})x^{n-1} + \ldots + (a_1 \pm b_1)x + (a_0 \pm b_0)$

6.4 Multiplying and Dividing Ponomials

Key Point

Multiplying monomials involves multiplying coefficients and adding exponents with the same base, while dividing entails dividing coefficients and subtracting exponents with the same base.

Formula To Remember

1 *Multiplying Monomials* $(ax^m) \times (bx^n) = (a \times b)x^{(m+n)}$

2 *Dividing Monomials* $\frac{ax^m}{bx^n} = \left(\frac{a}{b}\right)x^{(m-n)}$

6.5 Multiplying a Polynomial and a Monomial

Key Point

The Distributive Property is expressed as:

$$a(b+c) = ab+ac \quad \text{and} \quad a(b-c) = ab-ac,$$

where a, b, and c are variables or numbers.

Formula To Remember

1 *Distributive Property* $a(b+c) = ab+ac$ and $a(b-c) = ab-ac$, where a, b, c are algebraic expressions.

6.6 Multiplying Binomials

Key Point

The FOIL method for multiplying two binomials involves:

First: Multiply the first terms in each binomial.

Outer: Multiply the outer terms in the expression.

Inner: Multiply the inner terms.

Last: Multiply the last terms in each binomial.

Formula To Remember

1 *FOIL Method* For any binomials $(ax+b)(cx+d)$, apply the FOIL method:

$$(ax+b)(cx+d) = (ax)(cx)+(ax)(d)+(b)(cx)+(b)(d) = acx^2+(ad+bc)x+bd.$$

6.7 Factoring Trinomials

Key Point

To factor trinomials, you can use following methods:

- Using the 'FOIL' method, e.g., $(x+a)(x+b) = x^2 + (b+a)x + ab$.
- Using the 'Difference of Squares', e.g., $a^2 - b^2 = (a+b)(a-b)$.
- Using the 'Reverse FOIL' method, e.g., $x^2 + (b+a)x + ab = (x+a)(x+b)$.

Formula To Remember

1. *Difference of Squares* $a^2 - b^2 = (a+b)(a-b)$
2. *Reverse FOIL Method* $x^2 + (a+b)x + ab = (x+a)(x+b)$

6.8 Choosing a Factoring Method for Polynomials

Key Point

Factoring methods for common polynomial types:

- *Monomial*: Identify common factors.
- *Binomial*: Check for difference of squares or sum/difference of cubes.
- *Trinomial*: Use trial and error or the FOIL method.
- *Four terms*: Apply factoring by grouping.
- *General polynomials*: Look for common factors, then explore special identities (e.g., difference of squares, sum/difference of cubes).

Formula To Remember

1. *Sum/Difference of Cubes* $a^3 + b^3 = (a+b)(a^2 - ab + b^2)$, $a^3 - b^3 = (a-b)(a^2 + ab + b^2)$
2. *Factoring Trinomials (AC Method)* To factor $ax^2 + bx + c$, find two numbers that multiply to ac and add up to b (if they exist).

6.9 Factoring by Greatest Common Factor

Key Point

To find the Greatest Common Factor (GCF) of a polynomial:

- Determine the GCF of all terms.
- Express each term as a product of the GCF and a remaining factor.
- Extract the GCF from the polynomial using the Distributive Property.

6.10 Factors and Greatest Common Factors

Key Point

The GCF of a set of numbers is the largest number that divides all the numbers in the set without leaving a remainder.

Key Point

There are several common methods to find the GCF of a set of numbers, including listing factors, prime factorization, and using the division method.

6.11 Operations with Polynomials

Key Point

Polynomials can be added, subtracted, multiplied, or divided. Many of these operations involve the use of the Distributive Property.

Key Point

Subtraction of polynomials is similar to addition. However, every term of the subtracted polynomial changes its sign.

Key Point

To multiply polynomials, every term of the first polynomial is multiplied by every term from the second polynomial.

6.12 Even and Odd Functions

Key Point

Functions can be categorized as even or odd based on symmetry properties:

- *Even functions* exhibit y-axis symmetry, characterized by $f(-x) = f(x)$.
- *Odd functions* demonstrate origin symmetry, indicated by $f(-x) = -f(x)$.

Formula To Remember

1. *Even Function* ☞ $f(-x) = f(x)$ for all x in the domain.
2. *Odd Function* ☞ $f(-x) = -f(x)$ for all x in the domain.

6.13 End Behavior of Polynomial Functions

Key Point

The degree (n) and the leading coefficient (a_n) of a polynomial function $P(x) = a_n x^n + a_{n-1}x^{n-1} + \cdots + a_1 x + a_0$ govern the end behavior of the function.

Key Point

The easiest way to visualize the end behavior of a polynomial is to sketch a graph. However, you should be able to predict the end behavior without sketching, based solely on the degree and sign of the leading coefficient.

Formula To Remember

1. *End Behavior of Polynomial Functions with Even Degree and Positive Coefficient* 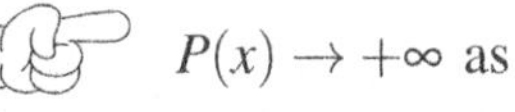 $P(x) \to +\infty$ as $x \to \pm\infty$
2. *End Behavior of Polynomial Functions with Even Degree and Negative Coefficient*  $P(x) \to -\infty$ as $x \to \pm\infty$
3. *End Behavior of Polynomial Functions with Odd Degree and Positive Coefficient* 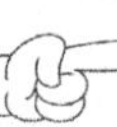 $P(x) \to +\infty$ as $x \to +\infty$ and $P(x) \to -\infty$ as $x \to -\infty$
4. *End Behavior of Polynomial Functions with Odd Degree and Negative Coefficient*  $P(x) \to -\infty$ as $x \to +\infty$ and $P(x) \to +\infty$ as $x \to -\infty$

6.14 Remainder and Factor Theorems

Key Point

Remainder Theorem states that if a polynomial $P(x)$ is divided by $x-a$, then the remainder is $P(a)$.

Key Point

Factor Theorem states that a polynomial $P(x)$ has a factor $x-c$ if and only if $P(c)=0$.

Formula To Remember

1. *Remainder Theorem* If $P(x)$ is divided by $x-a$, then the remainder is $P(a)$.
2. *Factor Theorem* $x-c$ is a factor of $P(x)$ if and only if $P(c)=0$.

6.15 Polynomial Division (Long Division)

Key Point

Long division of polynomials: Arrange both the dividend and the divisor in descending order of their degrees. Ensure that you include terms for any missing powers of x in the dividend with 0 coefficients.

6.16 Polynomial Division (Synthetic Division)

Key Point

In a divisor of the form $x-a$, the root is a. For a divisor of the form $x+a$, the root is $-a$.

Key Point

It is critical to ensure that the polynomial is written in standard form, and that a term is included for any missing degrees with a 0 coefficient.

6.17 Finding Zeros of Polynomials

Key Point

The number of zeros of a polynomial is up to its degree, and various methods like direct solving, the quadratic formula, factoring, the Remainder Theorem, or synthetic division are used to find these zeros.

Formula To Remember

1. *Zero of a Linear Polynomial* For $ax+b=c$, the zero is $x=\frac{c-b}{a}$.
2. *Zeros of a Quadratic Polynomial* For $ax^2+bx+c=0$, use $x=\frac{-b\pm\sqrt{b^2-4ac}}{2a}$.

6.18 Polynomial Identities

Key Point

Polynomial identities describe relationships between algebraic terms with variables, such as:

- $(x+y)^2=x^2+2xy+y^2$,
- $(x-y)^2=x^2-2xy+y^2$,
- $(x+y)(x-y)=x^2-y^2$.
- $(x+a)(x+b)=x^2+(a+b)x+ab$,
- $(x+y)^3=x^3+3x^2y+3xy^2+y^3=x^3+y^3+3xy(x+y)$,
- $(x-y)^3=x^3-3x^2y+3xy^2-y^3=x^3-y^3-3xy(x-y)$,
- $x^3+y^3=(x+y)\left(x^2-xy+y^2\right)$,
- $x^3-y^3=(x-y)\left(x^2+xy+y^2\right)$.

Formula To Remember

1. *Binomial Square Identities* $(x \pm y)^2 = x^2 \pm 2xy + y^2$
2. *Difference of Squares* $(x+y)(x-y) = x^2 - y^2$
3. *Binomial Cubes* $(x+y)^3 = x^3 + 3x^2y + 3xy^2 + y^3$, $(x-y)^3 = x^3 - 3x^2y + 3xy^2 - y^3$
4. *Sum and Difference of Cubes* $x^3 + y^3 = (x+y)(x^2 - xy + y^2)$, $x^3 - y^3 = (x-y)(x^2 + xy + y^2)$
5. *Other Identities* $(x+a)(x+b) = x^2 + (a+b)x + ab$ and $(x+y+z)^2 = x^2 + y^2 + z^2 + 2xy + 2yz + 2zx$

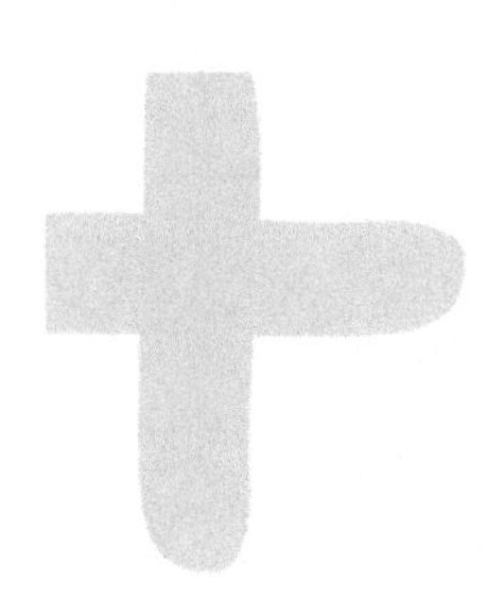

7. Functions Operations

7.1 Function Notation

Key Point

Functions in math link each input to a unique output, represented concisely by notations like $f(x)$, where 'f' is the function and 'x' the input.

Formula To Remember

1 *Evaluating a Function* To find $f(a)$, replace x in $f(x)$ with a: $f(a) = f(x)|_{x=a}$.

7.2 Adding and Subtracting Functions

Key Point

Adding or subtracting functions $f(x)$ and $g(x)$ yields

$$(f \pm g)(x) = f(x) \pm g(x),$$

combining them like algebraic expressions.

Formula To Remember

1. *Addition of Functions* $(f+g)(x)=f(x)+g(x)$
2. *Subtraction of Functions* $(f-g)(x)=f(x)-g(x)$

7.3 Multiplying and Dividing Functions

Key Point

For functions $f(x)$ and $g(x)$, multiplication and division create the new functions $(f\cdot g)(x)=f(x)\cdot g(x)$ and $\left(\frac{f}{g}\right)(x)=\frac{f(x)}{g(x)}$, considering $g(x)\neq 0$.

Formula To Remember

1. *Multiplication of Functions* $(f\cdot g)(x)=f(x)\cdot g(x)$
2. *Division of Functions* $\left(\frac{f}{g}\right)(x)=\frac{f(x)}{g(x)},\ g(x)\neq 0$

7.4 Composition of Functions

Key Point

The notation for the composition of two functions, f and g, is $(f\circ g)(x)$ or $f(g(x))$.

Formula To Remember

1. *Composition of Functions* $(f\circ g)(x)=f(g(x))$, which is not necessarily equal to $g(f(x))$.
2. *Evaluating Compositions* Given $f(x)$ and $g(x)$, for $(f\circ g)(a)$ substitute $g(a)$ in $f(x)$; for $(g\circ f)(a)$ substitute $f(a)$ in $g(x)$.

7.5 Writing Functions

Key Point

A function represents a special relationship between two variables, characterized by inputs (domain), outputs (range), and the rule (function rule) that specifies how inputs transform into outputs.

7.6 Parent Functions

Key Point

A parent function is the simplest form of a type of function, from which more complex functions can be derived through transformations

Key Point

Function $f(x)$ transformations with constant $k > 0$ include vertical $(f(x) \pm k)$, horizontal $(f(x \pm k))$, scaling $(kf(x), f(kx))$, and reflection $(-f(x), f(-x))$, preserving core properties except in reflections.

Formula To Remember

1. *Function Transformation: Up/Down* ☞ $f(x) \pm k$ (Add or subtract k to move up or down)
2. *Function Transformation: Left/Right* ☞ $f(x \pm k)$ (Replace x with $x \pm k$ to move left or right)
3. *Function Transformation: Narrower/Wider* ☞ $f(kx)$ (Replace x with kx to make the graph narrower for $k > 1$ or wider for $0 < k < 1$)
4. *Function Transformation: Stretch/Shrink* ☞ $kf(x)$ (Multiply the function by k to stretch for $k > 1$ or shrink for $0 < k < 1$)
5. *Function Transformation: Reflection* ☞ $-f(x)$ or $f(-x)$ (Multiply by -1 to reflect over the x-axis or replace x with $-x$ to reflect over the y-axis)

7.7 Function Inverses

Key Point

Function f and its inverse f^{-1} undo each other: $f(f^{-1}(y)) = y$ and $f^{-1}(f(x)) = x$.

Formula To Remember

1. *Definition of an Inverse Function* ☞ $f(f^{-1}(y)) = y$ and $f^{-1}(f(x)) = x$

7.8 Inverse Variation

Key Point

Inverse variation is characterized by the product of two variables remaining constant. This constant value is symbolized by k.

Formula To Remember

1 *General Form of Inverse Variation* $xy = k$ or $y = \frac{k}{x}$ where k is a non-zero constant, and $x \neq 0$, $y \neq 0$.

7.9 Graphing Functions

Key Point

A linear function can be expressed in the form $f(x) = ax + b$; its graph is a straight line in the coordinate plane.

Key Point

A quadratic function can be expressed in the form $f(x) = ax^2 + bx + c$; its graph is a parabola in the coordinate plane.

7.10 Domain and Range of Function

Key Point

The domain of a function $f(x)$ is the set of all possible input values (typically x values), and the range of $f(x)$ is the set of all corresponding output values (usually y values).

Key Point

The domain of a function varies: Polynomials and Exponentials include all real numbers; Square Roots $\sqrt{f(x)}$ require $f(x) \geq 0$; Logarithms $\log(f(x))$ need $f(x) > 0$; Rationals $\frac{f(x)}{g(x)}$ exclude $g(x) = 0$; Piecewise functions combine individual domains; and graph analysis can also determine domains.

Formula To Remember

1. *Domain of a Polynomial Function* ☞ All real numbers
2. *Domain of a Square Root Function* ☞ $f(x) \geq 0$ for $y = \sqrt{f(x)}$
3. *Domain of an Exponential Function* ☞ All real numbers
4. *Domain of a Logarithmic Function* ☞ $f(x) > 0$ for $y = \log(f(x))$
5. *Domain of a Rational Function* ☞ $\mathbb{R} - \{x|g(x) = 0\}$ for $y = \frac{f(x)}{g(x)}$

7.11 Piecewise Function

Key Point

Piecewise functions are divided into segments within their domains, each governed by a distinct formula.

Formula To Remember

1. *Absolute Value Function as a Piecewise Function* ☞ $f(x) = |x| = \begin{cases} x & \text{if } x \geq 0 \\ -x & \text{if } x < 0 \end{cases}$

7.12 Positive, Negative, Increasing, and Decreasing Functions

Key Point

In interval I, positive functions have graphs above the x-axis, while negative functions have graphs below it.

Key Point

A function is *increasing* if its output rises or remains constant as x increases, and *decreasing* if its output falls or remains constant.

Formula To Remember

1. *Positive Function* ☞ $f(x) > 0$ for every x in interval I.
2. *Negative Function* ☞ $f(x) < 0$ for every x in interval I.
3. *Increasing Function* ☞ $f(x_1) \leq f(x_2)$ for any $x_1 < x_2$ in interval I.
4. *Decreasing Function* ☞ $f(x_1) \geq f(x_2)$ for any $x_1 < x_2$ in interval I.

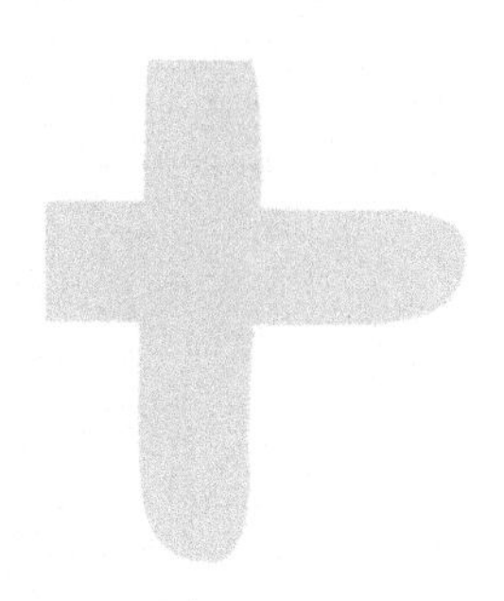

8. Exponential Functions

8.1 Exponential Function

Key Point

- Exponential functions $f(x) = a^x$ demonstrate growth for $a > 1$ (Exponential Growth) and decay for $0 < a < 1$ (Exponential Decay).
- These functions share a consistent Domain: $\mathbb{R}$, Range: $(0, +\infty)$, and y-intercept: 1.

Formula To Remember

1. *Definition of an Exponential Function* $f(x) = a^x$ where $a > 0$ and $a \neq 1$.
2. *Exponential Growth* If $a > 1$, the function $f(x) = a^x$ represents exponential growth.
3. *Exponential Decay* If $0 < a < 1$, the function $f(x) = a^x$ represents exponential decay.
4. *Domain and Range* Domain of $f(x) = a^x$ is $\mathbb{R}$, range is $(0, +\infty)$.
5. *y-Intercept of an Exponential Function* The y-intercept of $f(x) = a^x$ is $(0, 1)$.

8.2 Linear, Quadratic, and Exponential Models

Key Point

A linear function always results in a straight-line graph, a quadratic function results in a parabolic graph, and an exponential function results in a curve.

Formula To Remember

1. *Linear Function Form* ☞ $y = mx + b$ where m is the slope and b is the y-intercept.
2. *Quadratic Function Form* ☞ $y = ax^2 + bx + c$ where $a \neq 0$.
3. *Exponential Function Form* ☞ $y = a^x$ where $a > 0$ and $a \neq 1$.

8.3 Linear vs Exponential Growth

Key Point

Linear growth shows a constant y increase for uniform x increments (straight-line graph), while exponential growth multiplies y by a constant factor for each x increment (curved graph).

Formula To Remember

1. *Linear Growth Equation* ☞ $y = mx + b$ where m is the constant rate of change and b is the initial value.
2. *Exponential Growth Equation* ☞ $y = a \cdot r^x$ where a is the initial value and r is the base rate of increase, $r > 1$.

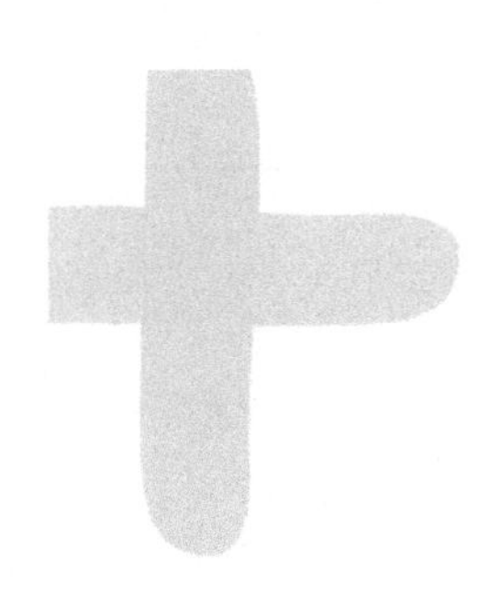

9. Logarithms

9.1 Evaluating Logarithms

Key Point

Key logarithmic rules are applicable under conditions where $a > 0$, $a \neq 1$, $M > 0$, $N > 0$, and k is a real number. These rules are:

- Rule 1: Product Rule - $\log_a(M \cdot N) = \log_a M + \log_a N$
- Rule 2: Quotient Rule - $\log_a\left(\frac{M}{N}\right) = \log_a M - \log_a N$
- Rule 3: Power Rule - $\log_a M^k = k\log_a M$
- Rule 4: Base Identity - $\log_a a = 1$
- Rule 5: Logarithm of One - $\log_a 1 = 0$
- Rule 6: Inverse Property - $a^{\log_a M} = M$

Formula To Remember

1. *Logarithm Definition* $\log_b y = x \Leftrightarrow y = b^x$ where $b > 0$, $b \neq 1$, and $y > 0$.
2. *Power Rule* $\log_a M^k = k\log_a M$ where $M > 0$, k is a real number.
3. *Base Identity* $\log_a a = 1$ where $a > 0$, $a \neq 1$.
4. *Logarithm of One* $\log_a 1 = 0$ where $a > 0$, $a \neq 1$.

9.2 Properties of Logarithms

Key Point

Properties of Logarithms:

$a^{\log_a b} = b$

$\log_a 1 = 0$

$\log_a a = 1$

$\log_a (x \cdot y) = \log_a x + \log_a y$

$\log_a \frac{x}{y} = \log_a x - \log_a y$

$\log_a \frac{1}{x} = -\log_a x$

$\log_a x^p = p \log_a x$

$\log_{a^k} x = \frac{1}{k} \log_a x$, for $k \neq 0$

$\log_a x = \log_{a^c} x^c$

$\log_a x = \frac{1}{\log_x a}$

Formula To Remember

1. *Product Rule* ☞ $\log_a(M \cdot N) = \log_a M + \log_a N$ where $M > 0$, $N > 0$.
2. *Quotient Rule* ☞ $\log_a\left(\frac{M}{N}\right) = \log_a M - \log_a N$ where $M > 0$, $N > 0$.
3. *Inverse Property* ☞ $a^{\log_a M} = M$ where $a > 0$, $a \neq 1$, $M > 0$.
4. *Change of Base Rule* ☞ $\log_a x = \frac{1}{\log_x a}$
5. *Reciprocal Rule* ☞ $\log_a \frac{1}{x} = -\log_a x$

9.3 Natural Logarithms

Key Point

A natural logarithm uses the base e (approximately 2.71) and is expressed as $\ln x$ or $\log_e x$, representing the logarithm of x to the base e.

Formula To Remember

1. *Definition of Natural Logarithm* ☞ $\ln x = \log_e x$, where $e \approx 2.71$.
2. *Exponential and Logarithm Relationship* ☞ $e^x = y \Leftrightarrow x = \ln(y)$.
3. *Natural Logarithm of e* ☞ $\ln(e) = 1$.

9.4 Solving Logarithmic Equations

Key Point

Solve logarithmic equations by isolating the variable, converting them to exponential form, combining logarithms, and verifying the validity of solutions.

Formula To Remember

1. *Converting Logarithmic to Exponential Form* ☞ $y = \log_b(x)$ if and only if $x = b^y$.
2. *Logarithmic Equation with Equal Bases* ☞ If $\log_b(m) = \log_b(n)$ then $m = n$.

10. Radical Expressions

10.1 Simplifying Radical Expressions

Key Point

Simplify radicals by:

- Exponent manipulation: $\sqrt[n]{x^a} = x^{\frac{a}{n}}$.
- Multiplication: $\sqrt[n]{xy} = x^{\frac{1}{n}} y^{\frac{1}{n}}$.
- Division: $\sqrt[n]{\frac{x}{y}} = \frac{x^{\frac{1}{n}}}{y^{\frac{1}{n}}}$.
- Product of radicals: $\sqrt[n]{x} \times \sqrt[n]{y} = \sqrt[n]{xy}$.

Formula To Remember

1. *Radical Exponent Conversion* ☞ $\sqrt[n]{x^a} = x^{\frac{a}{n}}$
2. *Multiplication of Radicals* ☞ $\sqrt[n]{xy} = \sqrt[n]{x}\sqrt[n]{y}$
3. *Simplification of Perfect Powers* ☞ $\sqrt[2k]{a^{2k}} = |a|$ and $\sqrt[2k+1]{a^{2k+1}} = a$, where $k \in \mathbb{N}$.

10.2 Simplifying Radical Expressions Involving Fractions

Key Point

The principal square root of a number is not negative; square roots of negative numbers are undefined.

Formula To Remember

1 *Division of Radicals* $\sqrt[n]{\frac{a}{b}} = \frac{\sqrt[n]{a}}{\sqrt[n]{b}}$ where a, b are non-negative numbers and $b \neq 0$.

10.3 Multiplying Radical Expressions

Key Point

The process of multiplying radical expressions typically involves:

1. Multiplying the coefficients outside the radicals.
2. Multiplying the contents inside the radicals.
3. Simplifying the resulting expression, as needed.

Formula To Remember

1 *Product of Radicals* $k_1\sqrt[n]{a} \times k_2\sqrt[n]{b} = k_1k_2\sqrt[n]{ab}$, where $a, b \geq 0$.

10.4 Adding and Subtracting Radical Expressions

Key Point

When working with radical expressions, only terms with similar radicals can be combined, while "unlik" radical terms cannot be added or subtracted.

Formula To Remember

 *Simplifying Like Radicals* $a\sqrt[n]{b} + c\sqrt[n]{b} = (a+c)\sqrt[n]{b}$ where $b \geq 0$.

10.5 Domain and Range of Radical Functions

Key Point

Domain and range of radical functions:

Domain: For $y = \sqrt{f(x)}$, it includes x values where $f(x) \geq 0$.

Range: In $y = c\sqrt{f(x)} + k$, the range is $y \geq k$ if $c \geq 0$ and $y \leq k$ if $c \leq 0$.

Formula To Remember

1. *Domain of a Radical Function* For $y = c\sqrt{f(x)} + k$, the domain includes all x such that $f(x) \geq 0$.
2. *Range of a Radical Function* For $y = c\sqrt{f(x)} + k$, the range is: $y \geq k$ if $c > 0$ and $y \leq k$ if $c < 0$.

10.6 Radical Equations

Key Point

Solve radical equations by isolating the radical, squaring both sides, solving, and checking for extraneous solutions.

10.7 Solving Radical Inequalities

Key Point

When dealing with a radical expression with an even index, ensure that the result is non-negative.

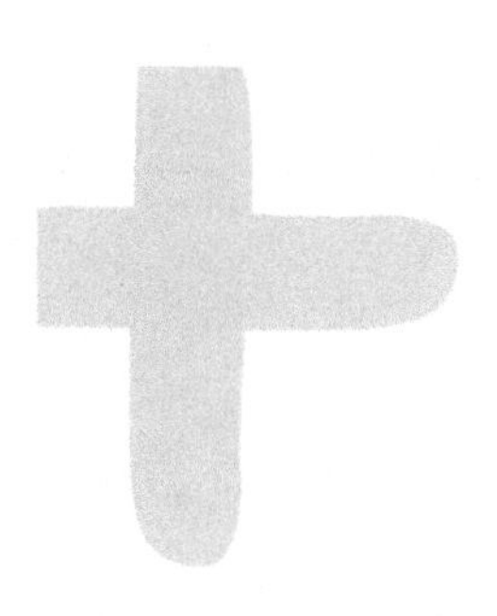

11. Rational and Irrational Expressions

11.1 Rational and Irrational Numbers

Key Point

Rational numbers, fractions $\frac{p}{q}$ with integer p and nonzero q, include natural and arithmetic numbers, and finite decimals. Also, There are infinite fractions between two rational numbers.

Formula To Remember

1. *Rational Numbers Definition* ☞ A number expressible as $\frac{p}{q}$ with $p, q \in \mathbb{Z}$ and $q \neq 0$.
2. *Irrational Numbers Definition* ☞ A real number that cannot be written as $\frac{p}{q}$ with $p, q \in \mathbb{Z}$ and $q \neq 0$.

11.2 Simplifying Rational Expressions

Key Point

Simplify rational expressions by dividing numerator and denominator by their Greatest Common Factor (GCF).

11.3 Graphing Rational Expressions

Key Point

To graph a rational function, identify and plot vertical asymptotes where the denominator equals zero, horizontal or oblique asymptotes based on polynomial degrees, and intercepts, then sketch the curve considering these features.

11.4 Multiplying Rational Expressions

Key Point

To multiply rational expressions, multiply the numerators to form the new numerator and the denominators to form the new denominator. The resulting expression is the product of the two rational expressions.

Formula To Remember

1 *Multiplication of Rational Expressions* ☞ $\frac{a}{b} \times \frac{c}{d} = \frac{a \times c}{b \times d}$, then factor, cancel common factors and simplify.

11.5 Dividing Rational Expressions

Key Point

Divide rational expressions by keeping the first, changing division to multiplication, flipping the second, and then multiplying across numerators and denominators.

Formula To Remember

1 *Division of Rational Expressions* ☞ $\frac{a}{b} \div \frac{c}{d} = \frac{a}{b} \times \frac{d}{c} = \frac{a \times d}{b \times c}$, then factor, cancel common factors and simplify.

11.6 Adding and Subtracting Rational Expressions

Key Point

To add or subtract rational expressions, find the Least Common Denominator (LCD), rewrite each expression with it, then add or subtract the numerators and simplify the result.

11.7 Rational Equations

Key Point

The common denominator method involves finding a common denominator for all fractions and equating the numerators, leading to a linear or quadratic equation.

Key Point

Cross-multiplication equates the product of the numerator of one fraction with the denominator of another, and is effective for equations of the form $\frac{a}{b} = \frac{c}{d}$.

Formula To Remember

1. *Common Denominator Method* ☞ $\frac{a}{b} = \frac{c}{d} \Rightarrow \frac{ad}{bd} = \frac{cb}{bd}$
2. *Cross-Multiplication* ☞ $\frac{a}{b} = \frac{c}{d} \Rightarrow ad = bc$

11.8 Simplifying Complex Fractions

Key Point

Simplifying a complex fraction, where either numerator or denominator is a fraction, involves converting to improper fractions, separating into division, and applying fraction division rules.

Formula To Remember

1. *Converting to Improper Fractions* ☞ Mixed number $a\frac{b}{c}$ to improper fraction $\frac{ac+b}{c}$.
2. *Division of Fractions (Keep, Change, Flip)* ☞ $\frac{a}{b} \div \frac{c}{d} = \frac{a}{b} \times \frac{d}{c} = \frac{ad}{bc}$.

11.9 Maximum and Minimum Points

Key Point

Functions may have multiple local extrema but only one absolute maximum or minimum.

11.10 Solving Rational Inequalities

Key Point

Solve rational inequalities by setting them to zero, finding critical points where numerator or denominator equals zero, testing intervals around these points, and using graphing as needed.

Formula To Remember

1. *General Form of a Rational Inequality* $\frac{P(x)}{Q(x)} < 0$ or $\frac{P(x)}{Q(x)} \leq 0$ or $\frac{P(x)}{Q(x)} > 0$ or $\frac{P(x)}{Q(x)} \geq 0$ where $P(x)$ and $Q(x)$ are polynomials, and $Q(x) \neq 0$.
2. *Identifying Critical Points* Solve $P(x) = 0$ and $Q(x) = 0$ for x to find the critical points.

11.11 Irrational Functions

Key Point

The domain of an irrational function depends on the radical's index and the properties of the function within the radical.

Formula To Remember

1. *General Form of an Irrational Function* $f(x) = \sqrt[n]{(g(x))^m}$ or $f(x) = (g(x))^{\frac{m}{n}}$
2. *Domain of an Irrational Function (Odd Index)* For odd n, the domain of $f(x)$ is the same as the domain of $g(x)$.
3. *Domain of an Irrational Function (Even Index)* For even n and any integer m, the domain of $f(x)$ is: $\{x \in \text{Domain of } g \mid (g(x))^m \geq 0\}$.

11.12 Direct, Inverse, Joint, and Combined Variation

Key Point

Types of variations: direct ($y = cx$), inverse ($y = \frac{c}{x}$), joint ($y = cxz$), and combined ($y = c\left(\frac{x}{z}\right)$).

Formula To Remember

1. *Direct Variation* ☞ $y = kx$ where k is the constant of variation.
2. *Inverse Variation* ☞ $y = \frac{k}{x}$ where k is the constant of variation.
3. *Joint Variation* ☞ $y = kxz$ where k is the constant of variation.
4. *Combined Variation* ☞ $y = k\left(\frac{x}{z}\right)$ where k is the constant of variation.

12. Trigonometric Functions

12.1 Angles of Rotation

Key Point

An angle is said to be in standard position when the initial side is along the positive x-axis and its vertex (endpoint) is at the origin.

Key Point

The reference angle, denoted by θ_{ref}, is the smallest angle formed by the terminal side of the given angle with the x-axis.

Formula To Remember

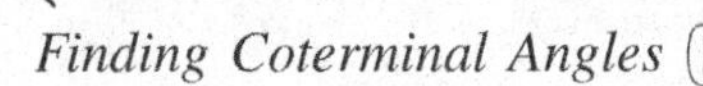

1 *Finding Coterminal Angles* $\theta' = \theta \pm 360°k$ or $\theta' = \theta \pm 2k\pi$ for any integer k

12.2 Angles and Angle Measure

Key Point

To convert an angle from degrees to radians, use the following formula:

$$\text{Radian} = \text{Degree} \times \frac{\pi}{180}.$$

To convert an angle from radians to degrees, use this formula:

$$\text{Degree} = \text{Radian} \times \frac{180}{\pi}.$$

Formula To Remember

1. *Converting Degrees to Radians* ☞ $\text{Radian} = \text{Degree} \times \frac{\pi}{180}$
2. *Converting Radians to Degrees* ☞ $\text{Degree} = \text{Radian} \times \frac{180}{\pi}$

12.3 Right-Triangle Trigonometry

Key Point

In a right triangle with one angle θ:

- The side opposite θ is known as the "opposite side."
- The side adjacent to θ is known as the "adjacent side."
- The longest side is called the "hypotenuse."

Key Point

The Pythagorean Theorem states that in a right triangle, the sum of the squares of the legs (a and b) equals the square of the hypotenuse (c), expressed as $c^2 = a^2 + b^2$.

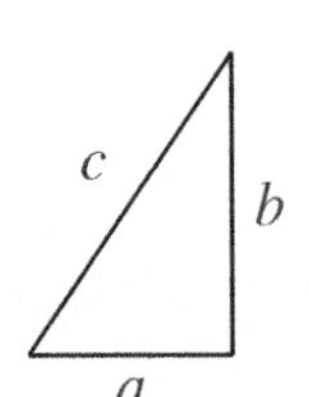

Formula To Remember

1. *Pythagorean Theorem* ☞ $c^2 = a^2 + b^2$ where c is the hypotenuse and a, b are the legs of the right triangle.

12.4 Trigonometric Ratios

Key Point

In a right triangle, the trigonometric ratios are:

$$\sin(\theta) = \frac{\text{opposite}}{\text{hypotenuse}}, \quad \cos(\theta) = \frac{\text{adjacent}}{\text{hypotenuse}}, \quad \text{and} \quad \tan(\theta) = \frac{\text{opposite}}{\text{adjacent}}.$$

Formula To Remember

1. *Sine of an angle* ☞ $\sin(\theta) = \frac{\text{opposite}}{\text{hypotenuse}}$
2. *Cosine of an angle* ☞ $\cos(\theta) = \frac{\text{adjacent}}{\text{hypotenuse}}$
3. *Tangent of an angle* ☞ $\tan(\theta) = \frac{\text{opposite}}{\text{adjacent}}$

12.5 Function Values and Ratios of Special and General Angles

Key Point

The table below shows exact values for $\sin(\theta)$, $\cos(\theta)$, and $\tan(\theta)$ at special angles, aiding in quick calculation and generalization to any angle.

θ	0°	30°	45°	60°	90°
$\sin(\theta)$	0	$\frac{1}{2}$	$\frac{\sqrt{2}}{2}$	$\frac{\sqrt{3}}{2}$	1
$\cos(\theta)$	1	$\frac{\sqrt{3}}{2}$	$\frac{\sqrt{2}}{2}$	$\frac{1}{2}$	0
$\tan(\theta)$	0	$\frac{\sqrt{3}}{3}$	1	$\sqrt{3}$	undefined

Key Point

Practical properties:

$$\sin(\theta) = \cos(90^\circ - \theta), \quad \cos(\theta) = \sin(90^\circ - \theta),$$

$$\cos(-\theta) = \cos(\theta), \quad \text{and} \quad \sin(-\theta) = -\sin(\theta).$$

Formula To Remember

1. *Values at 0 Degrees* ☞ $\sin(0^\circ) = 0, \quad \cos(0^\circ) = 1, \quad \tan(0^\circ) = 0$
2. *Values at 30 Degrees* ☞ $\sin(30^\circ) = \frac{1}{2}, \quad \cos(30^\circ) = \frac{\sqrt{3}}{2}, \quad \tan(30^\circ) = \frac{1}{\sqrt{3}}$
3. *Values at 45 Degrees* ☞ $\sin(45^\circ) = \frac{\sqrt{2}}{2}, \quad \cos(45^\circ) = \frac{\sqrt{2}}{2}, \quad \tan(45^\circ) = 1$
4. *Values at 60 Degrees* ☞ $\sin(60^\circ) = \frac{\sqrt{3}}{2}, \quad \cos(60^\circ) = \frac{1}{2}, \quad \tan(60^\circ) = \sqrt{3}$
5. *Values at 90 Degrees* ☞ $\sin(90^\circ) = 1, \quad \cos(90^\circ) = 0, \quad \tan(90^\circ) = \text{"undefined"}$
6. *Complementary Angle Identities* ☞ $\sin(\theta) = \cos(90^\circ - \theta)$ and $\cos(\theta) = \sin(90^\circ - \theta)$
7. *Even-Odd Identities* ☞ $\cos(-\theta) = \cos(\theta)$ and $\sin(-\theta) = -\sin(\theta)$

12.6 Missing Sides and Angles of a Right Triangle

Key Point

In right triangles, unknown sides can be calculated using

$$\sin(\theta) = \frac{\text{opposite}}{\text{hypotenuse}}, \quad \cos(\theta) = \frac{\text{adjacent}}{\text{hypotenuse}}, \quad \text{and} \quad \tan(\theta) = \frac{\text{opposite}}{\text{adjacent}},$$

given one side length and a non-right angle.

12.7 The Reciprocal Trigonometric Functions

Key Point

The reciprocal trigonometric functions are defined for angle θ as follows:

if $\cos(\theta) \neq 0$, $\sec(\theta) = \frac{1}{\cos(\theta)}$,

if $\sin(\theta) \neq 0$, $\csc(\theta) = \frac{1}{\sin(\theta)}$,

if $\tan(\theta) \neq 0$, $\cot(\theta) = \frac{1}{\tan(\theta)}$.

Formula To Remember

1. *Secant Function* ☞ $\sec(\theta) = \frac{1}{\cos(\theta)}$, for $\cos(\theta) \neq 0$.
2. *Cosecant Function* ☞ $\csc(\theta) = \frac{1}{\sin(\theta)}$, for $\sin(\theta) \neq 0$.
3. *Cotangent Function* ☞ $\cot(\theta) = \frac{1}{\tan(\theta)}$, for $\tan(\theta) \neq 0$.

12.8 Co-functions

Key Point

The five key co-function identities for complementary angles are:

1. $\sin(90^\circ - x) = \cos(x)$,
2. $\cos(90^\circ - x) = \sin(x)$,
3. $\cot(90^\circ - x) = \tan(x)$,
4. $\sec(90^\circ - x) = \csc(x)$,
5. $\csc(90^\circ - x) = \sec(x)$.

Formula To Remember

1. *Sine Co-function Identity* ☞ $\sin(90^\circ - x) = \cos(x)$
2. *Cosine Co-function Identity* ☞ $\cos(90^\circ - x) = \sin(x)$
3. *Tangent Co-function Identity* ☞ $\tan(90^\circ - x) = \cot(x)$
4. *Cotangent Co-function Identity* ☞ $\cot(90^\circ - x) = \tan(x)$
5. *Secant Co-function Identity* ☞ $\sec(90^\circ - x) = \csc(x)$
6. *Cosecant Co-function Identity* ☞ $\csc(90^\circ - x) = \sec(x)$

12.9 Function Values from the Calculator

Key Point

Ensure you correctly select the mode (Degree or Radian), as per the angle you are dealing with, on your calculator.

Key Point

It is important to remember that calculator outputs are typically approximations, not exact values.

12.10 Reference Angles and the Calculator

Key Point

Quadrant characteristics:

1. First Quadrant (0° to 90°): All functions positive.
2. Second Quadrant (90° to 180°): sin positive, cos and tan negative.
3. Third Quadrant (180° to 270°): sin and cos negative, tan positive.
4. Fourth Quadrant (270° to 360°): cos positive, sin and tan negative.

Key Point

For reference angles:

1. First Quadrant: Same as the given angle.
2. Second Quadrant: $180^\circ - \theta$.
3. Third Quadrant: $\theta - 180^\circ$.
4. Fourth Quadrant: $360^\circ - \theta$.

Normalize angles outside 0° to 360° by adding/subtracting 360°.

Formula To Remember

1. *Quadrant Function Signs* First Quadrant: All functions positive. Second Quadrant: sin positive, cos and tan negative. Third Quadrant: sin and cos negative, tan positive. Fourth Quadrant: cos positive, sin and tan negative.

2. *Finding Reference Angles* First Quadrant: θ. Second Quadrant: $180^\circ - \theta$. Third Quadrant: $\theta - 180^\circ$. Fourth Quadrant: $360^\circ - \theta$.

12.11 Coterminal Angles and Reference Angles

Key Point

An angle is said to be Coterminal with another if they have the same terminal side.

Key Point

The Reference Angle is the smallest angle that you can make from the terminal side of an angle with the x-axis.

12.12 Evaluating Trigonometric Function

Key Point

Evaluating Trigonometric Functions:

1. Draw the terminal side of the angle.
2. Find the reference angle, the smallest angle from the terminal side to the x-axis.
3. Calculate the trigonometric function of the reference angle.

12.13 Pythagorean Identities

Key Point

The Pythagorean identity, $\sin^2\theta+\cos^2\theta=1$, is essential in trigonometry and leads to two derived identities:

- Dividing by $\sin^2\theta$: $1+\cot^2\theta=\csc^2\theta$.
- Dividing by $\cos^2\theta$: $\tan^2\theta+1=\sec^2\theta$.

Formula To Remember

1. *Fundamental Pythagorean Identity* ☞ $\sin^2\theta+\cos^2\theta=1$ for any angle θ.
2. *Derived Pythagorean Identity (by dividing by $\sin^2\theta$)* ☞ $1+\cot^2\theta=\csc^2\theta$
3. *Derived Pythagorean Identity (by dividing by $\cos^2\theta$)* ☞ $\tan^2\theta+1=\sec^2\theta$

12.14 The Unit Circle, Sine, and Cosine

Key Point

The unit circle is instrumental for understanding the sine and cosine functions. The y-coordinate of a point on the unit circle provides the value of $\sin\theta$, and the x-coordinate yields the value of $\cos\theta$.

Formula To Remember

1. *Unit Circle Equation* ☞ $x^2 + y^2 = 1$
2. *Sine from Unit Circle* ☞ $\sin\theta = y$-coordinate of point on the unit circle
3. *Cosine from Unit Circle* ☞ $\cos\theta = x$-coordinate of point on the unit circle

12.15 Arc Length and Sector Area

Area of a sector with a central angle θ (in degrees) $= \pi r^2 \left(\frac{\theta}{360}\right)$.

Key Point

Arc length of a sector with a central angle θ (in degrees) $= \pi r \left(\frac{\theta}{180}\right)$.

Formula To Remember

1. *Area of a Sector* ☞ $A = \pi r^2 \left(\frac{\theta}{360}\right)$ where r is the radius, and θ is the central angle in degrees.
2. *Arc Length* ☞ $L = \pi r \left(\frac{\theta}{180}\right)$ where r is the radius, and θ is the central angle in degrees.

13. Sequences and Series

13.1 Arithmetic Sequences

Key Point

To find any term in an arithmetic sequence, use the formula:

$$x_n = a + (n-1)d,$$

where a is the first term, d is the common difference, and n is the term number.

Formula To Remember

1 *n-th Term of an Arithmetic Sequence* $x_n = a + (n-1)d$ where a is the first term, d is the common difference, and n is the term number.

13.2 Geometric Sequences

Key Point

To find any term in a geometric sequence, use the formula:

$$x_n = ar^{(n-1)},$$

where n is the term number, x_n is the n-th term, a is the first term, and r is the common ratio.

 **Formula To Remember**

1. *General Form of a Geometric Sequence* $x_n = ar^{(n-1)}$ where x_n is the n-th term, a is the first term, and r is the common ratio.

13.3 Finite Arithmetic Series

Key Point

The sum of an arithmetic series is given by the formula:

$$S_n = \frac{1}{2}n(2a + d(n-1)),$$

where n is the number of terms, a is the first term, and d is the common difference. A more concise form of the formula is: $S_n = \frac{1}{2}n(a + x_n)$, where x_n is the n-th term.

Formula To Remember

1. *Sum of a Finite Arithmetic Series* $S_n = \frac{1}{2}n(2a + d(n-1))$ where n is the number of terms, a is the first term, and d is the common difference.
2. *Concise Sum of a Finite Arithmetic Series* $S_n = \frac{1}{2}n(a + x_n)$ where n is the number of terms, a is the first term, and x_n is the n-th term.

13.4 Finite Geometric Series

Key Point

The formula for the sum S_n of a finite geometric series is: $S_n = a\left(\frac{1-r^n}{1-r}\right)$, where n is the term count, a is the initial term, and r is the common ratio.

 **Formula To Remember**

1. *Sum of a Finite Geometric Series* $S_n = a\left(\frac{1-r^n}{1-r}\right)$ where n is the number of terms, a is the first term, and r is the common ratio.

13.5 Infinite Geometric Series

Key Point

Infinite Geometric Series:

- The sum is infinite when the absolute value of the ratio is greater than 1.
- Formula: $S = \frac{a}{1-r}$ with first term a and common ratio r where $|r| < 1$.

Formula To Remember

1 *Sum of an Infinite Geometric Series* $S = \frac{a}{1-r}$ where a is the first term and r is the common ratio ($|r| < 1$).

13.6 Pascal's Triangle

Key Point

In Pascal's Triangle, the n-th row (starting from 0) has $n+1$ entries. The k-th entry of this row is equal to $\binom{n}{k} = \dfrac{n!}{k!(n-k)!}$, where n is the row number, and k is the position of the entry within that row, starting from 0.

Formula To Remember

1 *Pascal's Triangle nth Entry* The k-th entry of the n-th row is $\binom{n}{k} = \frac{n!}{k!(n-k)!}$.

13.7 Binomial Theorem

Key Point

In the expansion of $(x+y)^n$, the exponents on x start with n and decrease, while the exponents on y start with 0 and increase. The powers on x and y always add up to n in each term.

Formula To Remember

1. *Binomial Theorem Formula* ☞ $(x+y)^n = \sum_{k=0}^{n} \binom{n}{k} x^{n-k} y^k$
2. *Binomial Coefficient* ☞ $\binom{n}{k} = \frac{n!}{k!(n-k)!}$
3. *k-th Term of a Binomial Expansion* ☞ The k-th term is given by $\binom{n}{k-1} x^{n-k+1} y^{k-1}$

13.8 Sigma Notation (Summation Notation)

Key Point

Summation notation is a powerful tool for working with sequences and series and makes calculations much easier.

Formula To Remember

1. *Definition of Sigma Notation* ☞ $\sum_{i=1}^{n} f(i) = f(1) + f(2) + \cdots + f(n)$ where i an indexing variable and $f(i)$ is a function of i.

13.9 Alternate Series

Key Point

The first term of an alternating series can be either positive or negative, depending on the index used to start the series.

Key Point

An alternating series is convergent if its terms approach 0, and each term is equal to or less than the preceding term.

Formula To Remember

1. *General Form of an Alternating Series* ☞ $\sum_{k=1}^{\infty} (-1)^k a_k$ where $a_k \geq 0$.
2. *Convergence of an Alternating Series* ☞ An alternating series converges if $0 \leq a_{n+1} \leq a_n$ for all $n \geq 1$ and $a_n \to 0$ as $n \to \infty$.

It is Time to Test Yourself

In the following, there are two complete College Algebra Tests. Once you've completed them, evaluate your performance by using the provided answer key.

- **Gather your supplies:** Ensure you have a pencil and a calculator ready before starting the test.
- **Question types:** There are two types of questions you'll encounter:

  1) **Multiple choice** questions where have four or more answer choices for these questions.

  2) **Grid-ins questions**: You'll need to write your answer in the provided box.

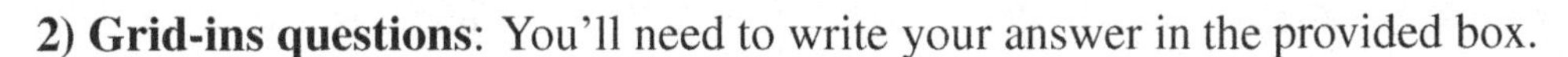

- **Don't be afraid to guess:** It's perfectly fine to make educated guesses. Remember, there are no penalties for wrong answers.
- **Review your work:** After completing the test, take some time to go over the answer key. It will help you identify where you made mistakes and areas that need improvement.
- **Stay calm and confident:** Believe in yourself and your abilities. You've got this!

14. Formula Sheet for All Topics

For your benefit we are providing all formulas again in one place. This is a quick reference guide to the formulas you need to remember for Algebra topics. We recommend that you review this formula sheet before taking the practice tests. This will help you recall and apply the formulas effectively.

Formulas For Chapter Fundamentals and Building Blocks

Order of Operations

1. *Order of Operations (PEMDAS)* Parentheses → Exponents → Multiplication/Division (left to right) → Addition/Subtraction (left to right).

Scientific Notation

1. *Writing in Scientific Notation* $m \times 10^n$ where $1 \leq m < 10$ and n is an integer.

Rules of Exponents

1. *Product of Powers* $x^m \times x^n = x^{(m+n)}$
2. *Quotient of Powers* $\frac{x^m}{x^n} = x^{(m-n)}$
3. *Power of a Power* $(x^m)^n = x^{mn}$
4. *Power of a Product* $(xy)^n = x^n y^n$
5. *Power of a Quotient* $\left(\frac{x}{y}\right)^n = \frac{x^n}{y^n}$

Formulas For Chapter Equations and Inequalities

Slope and Intercepts

1 *Slope of a Line Through Two Points* ☞ $m = \frac{y_2 - y_1}{x_2 - x_1}$ where $A(x_1, y_1)$ and $B(x_2, y_2)$ are two distinct points on the line.

2 *Slope-Intercept Form of a Line* ☞ $y = mx + b$ where m is the slope and $(0, b)$ is the y-intercept.

Using Intercepts

1 *x-Intercept* ☞ Set $y = 0$ in $ax + by = c$. Solve for x to get x-intercept $(x, 0)$.

2 *y-Intercept* ☞ Set $x = 0$ in $ax + by = c$. Solve for y to get y-intercept $(0, y)$.

Transforming Linear Functions

1 *Translation: Vertical Shift of a Line* ☞ $y = mx + (b \pm k)$ translates the line k units vertically.

2 *Rotation: Changing the Slope* ☞ $y = (m \pm k)x + b$ rotates the line around the point $(0, b)$, altering steepness.

3 *Reflection: Flipping Across the y-axis* ☞ $y = -mx + b$ reflects the line across the y-axis.

Solving Compound Inequalities

1 *Conjunction (AND)* ☞ Solve A and B separately, then find the intersection. For $A \cap B$, x must satisfy both A and B.

2 *Disjunction (OR)* ☞ Solve A and B separately, then find the union. For $A \cup B$, x satisfies either A or B.

3 *Graphing Compound Inequalities* ☞ Use open circles for exclusive ($<$, $>$) and closed circles for inclusive ($\leq$, $\geq$). Draw lines to show the range of x values that satisfy the inequality.

Solving Absolute Value Equations

1 *Standard Absolute Value Equation* ☞ $|ax + b| = c$, then we have $x = \frac{c-b}{a}$ and $x = \frac{-c-b}{a}$

Solving Absolute Value Inequalities

1. *Solving "$|ax+b| \leq c$"* ☞ $-c \leq ax+b \leq c$
2. *Solving "$|ax+b| \geq c$"* ☞ $ax+b \geq c$ or $ax+b \leq -c$

Solving Special Systems

1. *Condition for Parallel Lines* ☞ Two equations $y = m_1x + b_1$ and $y = m_2x + b_2$ are parallel (and have no solution) if $m_1 = m_2$ and $b_1 \neq b_2$.
2. *Condition for Overlapping Lines* ☞ Two equations $y = m_1x + b_1$ and $y = m_2x + b_2$ overlap (and have an infinite number of solutions) if $m_1 = m_2$ and $b_1 = b_2$.

Formulas For Chapter Quadratic Function

Solving a Quadratic Equation

1. *Standard Form of a Quadratic Equation* ☞ $ax^2 + bx + c = 0$ where $a \neq 0$.
2. *Discriminant* ☞ $D = b^2 - 4ac$
3. *Quadratic Formula* ☞ $x = \frac{-b \pm \sqrt{D}}{2a}$ where $a \neq 0$.
4. *Nature of Roots* ☞ If $D > 0$, two distinct real roots; if $D = 0$, one real root; if $D < 0$, no real roots.

Graphing Quadratic Functions

1. *Standard Form of a Quadratic Function* ☞ $y = ax^2 + bx + c$ where $a \neq 0$, b, and c are constants.
2. *Vertex Form of a Quadratic Function* ☞ $y = a(x-h)^2 + k$ $(a \neq 0)$ where (h,k) is the vertex and $x = h$ is the axis of symmetry.
3. *Vertex from Standard Form* ☞ Vertex (h,k) is found by $h = -\frac{b}{2a}$ and k by substituting $x = h$ into the standard form.

Axis of Symmetry of Quadratic Functions

1. *Axis of Symmetry (Standard Form)* ☞ $x = -\frac{b}{2a}$ for the quadratic function $y = ax^2 + bx + c$.
2. *Axis of Symmetry (Vertex Form)* ☞ $x = h$ for the quadratic function $y = a(x-h)^2 + k$.

Solving Quadratic Equations with Square Roots

1. *Solution for* $ax^2 = c$ ☞ $x = \pm\sqrt{\frac{c}{a}}$
2. *Solution for* $(ax+b)^2 = c$ ☞ $x = \frac{-b \pm \sqrt{c}}{a}$

Build Quadratics from Roots

1. *Quadratic Equation from Roots* ☞ If α and β are roots we have $x^2 - (\alpha+\beta)x + \alpha\beta = 0$

Factoring the Difference of Two Perfect Squares

1. *Difference of Two Perfect Squares* ☞ $a^2 - b^2 = (a-b)(a+b)$

Formulas For Chapter Complex Numbers

Adding and Subtracting Complex Numbers

1. *Addition of Complex Numbers* ☞ $(a+bi) + (c+di) = (a+c) + (b+d)i$
2. *Subtraction of Complex Numbers* ☞ $(a+bi) - (c+di) = (a-c) + (b-d)i$

Multiplying and Dividing Complex Numbers

1. *Multiplying Complex Numbers* ☞ $(a+bi)(c+di) = (ac-bd) + (ad+bc)i$
2. *Conjugate of a Complex Number* ☞ For $z = a+bi$ the conjugate, $\bar{z}$, is $a-bi$
3. *Dividing Complex Numbers* ☞ $\frac{a+bi}{c+di} = \frac{a+bi}{c+di} \times \frac{c-di}{c-di} = \frac{ac+bd}{c^2+d^2} + \frac{bc-ad}{c^2+d^2}i$

Rationalizing Imaginary Denominators

1. *Rationalization of a Complex Fraction* Multiply by the conjugate: $\frac{a+bi}{c+di} \times \frac{c-di}{c-di}$

Formulas For Chapter Matrices

Adding and Subtracting Matrices

1. *Matrix Addition and Subtraction* Given matrices A and B of the same dimensions $m \times n$, the sum or difference C is defined as: $C = A \pm B$ where $C_{ij} = A_{ij} \pm B_{ij}$ for all $1 \leq i \leq m$ and $1 \leq j \leq n$.

Matrix Multiplication

1. *Matrix Multiplication Criteria* The number of columns in the first matrix must be equal to the number of rows in the second matrix.
2. *Matrix Multiplication Process* If A is an $m \times n$ matrix and B is an $n \times p$ matrix, then $C = A \times B$ is an $m \times p$ matrix with entries $c_{ij} = a_{i1}b_{1j} + a_{i2}b_{2j} + \cdots + a_{in}b_{nj}$.

Finding Determinants of a Matrix

1. *Determinant of a 2×2 Matrix* For $A = \begin{bmatrix} a & b \\ c & d \end{bmatrix}$, $|A| = ad - bc$.
2. *Determinant of a 3×3 Matrix* For $A = \begin{bmatrix} a & b & c \\ d & e & f \\ g & h & i \end{bmatrix}$, $|A| = a(ei - fh) - b(di - fg) + c(dh - eg)$.

The Inverse of a Matrix

1. *Condition for the Existence of an Inverse* A matrix A has an inverse A^{-1} if it is square and non-singular (i.e., $\det(A) \neq 0$).
2. *Inverse of a 2×2 Matrix* For $A = \begin{bmatrix} a & b \\ c & d \end{bmatrix}$, its inverse $A^{-1} = \frac{1}{ad-bc} \begin{bmatrix} d & -b \\ -c & a \end{bmatrix}$.

Solving Linear Systems with Matrix Equations

1. *Writing System of Equations as Matrix Equation* $AX = B$ where A is the coefficient matrix, X is the variable matrix, and B is the constant matrix.
2. *Matrix Equation Solution Using Inverse* $X = A^{-1}B$, if $|A| \neq 0$.

Formulas For Chapter Polynomial Operations

Writing Polynomials in Standard Form

1. *Standard Form of a Polynomial* $f(x) = a_n x^n + a_{n-1} x^{n-1} + \cdots + a_1 x + a_0$ where each a_i is a coefficient and n is the degree.
2. *Degree of the Polynomial* The degree is the highest exponent of x.

Simplifying Polynomials

1. *Combining Like Terms* $ax^n y^m + bx^n y^m = (a+b)x^n y^m$

Adding and Subtracting Polynomials

1. *Adding and Subtracting Polynomials* $(a_n x^n + a_{n-1} x^{n-1} + \ldots + a_1 x + a_0) \pm (b_n x^n + b_{n-1} x^{n-1} + \ldots + b_1 x + b_0) = (a_n \pm b_n) x^n + (a_{n-1} \pm b_{n-1}) x^{n-1} + \ldots + (a_1 \pm b_1) x + (a_0 \pm b_0)$

Multiplying and Dividing Monomials

1. *Multiplying Monomials* $(ax^m) \times (bx^n) = (a \times b) x^{(m+n)}$
2. *Dividing Monomials* $\frac{ax^m}{bx^n} = \left(\frac{a}{b}\right) x^{(m-n)}$

Multiplying a Polynomial and a Monomial

1. *Distributive Property* $a(b+c) = ab + ac$ and $a(b-c) = ab - ac$, where a, b, c are algebraic expressions.

Multiplying Binomials

1. *FOIL Method* For any binomials $(ax+b)(cx+d)$, apply the FOIL method:

$$(ax+b)(cx+d) = (ax)(cx) + (ax)(d) + (b)(cx) + (b)(d) = acx^2 + (ad+bc)x + bd.$$

Factoring Trinomials

1. *Difference of Squares* $a^2 - b^2 = (a+b)(a-b)$
2. *Reverse FOIL Method* $x^2 + (a+b)x + ab = (x+a)(x+b)$

Choosing a Factoring Method for Polynomials

1. *Sum/Difference of Cubes* $a^3 + b^3 = (a+b)(a^2 - ab + b^2)$, $a^3 - b^3 = (a-b)(a^2 + ab + b^2)$
2. *Factoring Trinomials (AC Method)* To factor $ax^2 + bx + c$, find two numbers that multiply to ac and add up to b (if they exist).

Even and Odd Functions

1. *Even Function* $f(-x) = f(x)$ for all x in the domain.
2. *Odd Function* $f(-x) = -f(x)$ for all x in the domain.

End Behavior of Polynomial Functions

1. *End Behavior of Polynomial Functions with Even Degree and Positive Coefficient* $P(x) \to +\infty$ as $x \to \pm\infty$
2. *End Behavior of Polynomial Functions with Even Degree and Negative Coefficient* $P(x) \to -\infty$ as $x \to \pm\infty$
3. *End Behavior of Polynomial Functions with Odd Degree and Positive Coefficient* $P(x) \to +\infty$ as $x \to +\infty$ and $P(x) \to -\infty$ as $x \to -\infty$
4. *End Behavior of Polynomial Functions with Odd Degree and Negative Coefficient* $P(x) \to -\infty$ as $x \to +\infty$ and $P(x) \to +\infty$ as $x \to -\infty$

Remainder and Factor Theorems

1. *Remainder Theorem* If $P(x)$ is divided by $x - a$, then the remainder is $P(a)$.
2. *Factor Theorem* $x - c$ is a factor of $P(x)$ if and only if $P(c) = 0$.

Finding Zeros of Polynomials

1. *Zero of a Linear Polynomial* For $ax + b = c$, the zero is $x = \frac{c-b}{a}$.
2. *Zeros of a Quadratic Polynomial* For $ax^2 + bx + c = 0$, use $x = \frac{-b \pm \sqrt{b^2 - 4ac}}{2a}$.

Polynomial Identities

1. *Binomial Square Identities* $(x \pm y)^2 = x^2 \pm 2xy + y^2$
2. *Difference of Squares* $(x + y)(x - y) = x^2 - y^2$
3. *Binomial Cubes* $(x + y)^3 = x^3 + 3x^2y + 3xy^2 + y^3$, $(x - y)^3 = x^3 - 3x^2y + 3xy^2 - y^3$
4. *Sum and Difference of Cubes* $x^3 + y^3 = (x + y)(x^2 - xy + y^2)$, $x^3 - y^3 = (x - y)(x^2 + xy + y^2)$
5. *Other Identities* $(x + a)(x + b) = x^2 + (a + b)x + ab$ and $(x + y + z)^2 = x^2 + y^2 + z^2 + 2xy + 2yz + 2zx$

Formulas For Chapter Functions Operations

Function Notation

1. *Evaluating a Function* To find $f(a)$, replace x in $f(x)$ with a: $f(a) = f(x)|_{x=a}$.

Adding and Subtracting Functions

1. *Addition of Functions* $(f + g)(x) = f(x) + g(x)$
2. *Subtraction of Functions* $(f - g)(x) = f(x) - g(x)$

Multiplying and Dividing Functions

1. *Multiplication of Functions* ☞ $(f \cdot g)(x) = f(x) \cdot g(x)$
2. *Division of Functions* ☞ $\left(\frac{f}{g}\right)(x) = \frac{f(x)}{g(x)}$, $g(x) \neq 0$

Composition of Functions

1. *Composition of Functions* ☞ $(f \circ g)(x) = f(g(x))$, which is not necessarily equal to $g(f(x))$.
2. *Evaluating Compositions* ☞ Given $f(x)$ and $g(x)$, for $(f \circ g)(a)$ substitute $g(a)$ in $f(x)$; for $(g \circ f)(a)$ substitute $f(a)$ in $g(x)$.

Parent Functions

1. *Function Transformation: Up/Down* ☞ $f(x) \pm k$ (Add or subtract k to move up or down)
2. *Function Transformation: Left/Right* ☞ $f(x \pm k)$ (Replace x with $x \pm k$ to move left or right)
3. *Function Transformation: Narrower/Wider* ☞ $f(kx)$ (Replace x with kx to make the graph narrower for $k > 1$ or wider for $0 < k < 1$)
4. *Function Transformation: Stretch/Shrink* ☞ $kf(x)$ (Multiply the function by k to stretch for $k > 1$ or shrink for $0 < k < 1$)
5. *Function Transformation: Reflection* ☞ $-f(x)$ or $f(-x)$ (Multiply by -1 to reflect over the x-axis or replace x with $-x$ to reflect over the y-axis)

Function Inverses

1. *Definition of an Inverse Function* ☞ $f(f^{-1}(y)) = y$ and $f^{-1}(f(x)) = x$

Inverse Variation

1. *General Form of Inverse Variation* ☞ $xy = k$ or $y = \frac{k}{x}$ where k is a non-zero constant, and $x \neq 0$, $y \neq 0$.

Domain and Range of Function

1. *Domain of a Polynomial Function* ☞ All real numbers
2. *Domain of a Square Root Function* ☞ $f(x) \geq 0$ for $y = \sqrt{f(x)}$
3. *Domain of an Exponential Function* ☞ All real numbers
4. *Domain of a Logarithmic Function* ☞ $f(x) > 0$ for $y = \log(f(x))$
5. *Domain of a Rational Function* ☞ $\mathbb{R} - \{x|g(x) = 0\}$ for $y = \frac{f(x)}{g(x)}$

Piecewise Function

1. *Absolute Value Function as a Piecewise Function* ☞ $f(x) = |x| = \begin{cases} x & \text{if } x \geq 0 \\ -x & \text{if } x < 0 \end{cases}$

Positive, Negative, Increasing, and Decreasing Functions

1. *Positive Function* ☞ $f(x) > 0$ for every x in interval I.
2. *Negative Function* ☞ $f(x) < 0$ for every x in interval I.
3. *Increasing Function* ☞ $f(x_1) \leq f(x_2)$ for any $x_1 < x_2$ in interval I.
4. *Decreasing Function* ☞ $f(x_1) \geq f(x_2)$ for any $x_1 < x_2$ in interval I.

Formulas For Chapter Exponential Functions

Exponential Function

1. *Definition of an Exponential Function* ☞ $f(x) = a^x$ where $a > 0$ and $a \neq 1$.
2. *Exponential Growth* ☞ If $a > 1$, the function $f(x) = a^x$ represents exponential growth.
3. *Exponential Decay* ☞ If $0 < a < 1$, the function $f(x) = a^x$ represents exponential decay.
4. *Domain and Range* ☞ Domain of $f(x) = a^x$ is $\mathbb{R}$, range is $(0, +\infty)$.
5. *y-Intercept of an Exponential Function* ☞ The y-intercept of $f(x) = a^x$ is $(0, 1)$.

Linear, Quadratic, and Exponential Models

1. *Linear Function Form* ☞ $y = mx + b$ where m is the slope and b is the y-intercept.
2. *Quadratic Function Form* ☞ $y = ax^2 + bx + c$ where $a \neq 0$.
3. *Exponential Function Form* ☞ $y = a^x$ where $a > 0$ and $a \neq 1$.

Linear vs Exponential Growth

1. *Linear Growth Equation* ☞ $y = mx + b$ where m is the constant rate of change and b is the initial value.
2. *Exponential Growth Equation* ☞ $y = a \cdot r^x$ where a is the initial value and r is the base rate of increase, $r > 1$.

Formulas For Chapter Logarithms

Evaluating Logarithms

1. *Logarithm Definition* ☞ $\log_b y = x \Leftrightarrow y = b^x$ where $b > 0$, $b \neq 1$, and $y > 0$.
2. *Power Rule* ☞ $\log_a M^k = k \log_a M$ where $M > 0$, k is a real number.
3. *Base Identity* ☞ $\log_a a = 1$ where $a > 0$, $a \neq 1$.
4. *Logarithm of One* ☞ $\log_a 1 = 0$ where $a > 0$, $a \neq 1$.

Properties of Logarithms

1. *Product Rule* ☞ $\log_a(M \cdot N) = \log_a M + \log_a N$ where $M > 0$, $N > 0$.
2. *Quotient Rule* ☞ $\log_a\left(\frac{M}{N}\right) = \log_a M - \log_a N$ where $M > 0$, $N > 0$.
3. *Inverse Property* ☞ $a^{\log_a M} = M$ where $a > 0$, $a \neq 1$, $M > 0$.
4. *Change of Base Rule* ☞ $\log_a x = \frac{1}{\log_x a}$
5. *Reciprocal Rule* ☞ $\log_a \frac{1}{x} = -\log_a x$

Natural Logarithms

1. *Definition of Natural Logarithm* ☞ $\ln x = \log_e x$, where $e \approx 2.71$.
2. *Exponential and Logarithm Relationship* ☞ $e^x = y \Leftrightarrow x = \ln(y)$.
3. *Natural Logarithm of e* ☞ $\ln(e) = 1$.

Solving Logarithmic Equations

1. *Converting Logarithmic to Exponential Form* ☞ $y = \log_b(x)$ if and only if $x = b^y$.
2. *Logarithmic Equation with Equal Bases* ☞ If $\log_b(m) = \log_b(n)$ then $m = n$.

Formulas For Chapter Radical Expressions

Simplifying Radical Expressions

1. *Radical Exponent Conversion* ☞ $\sqrt[n]{x^a} = x^{\frac{a}{n}}$
2. *Multiplication of Radicals* ☞ $\sqrt[n]{xy} = \sqrt[n]{x}\sqrt[n]{y}$
3. *Simplification of Perfect Powers* ☞ $\sqrt[2k]{a^{2k}} = |a|$ and $\sqrt[2k+1]{a^{2k+1}} = a$, where $k \in \mathbb{N}$.

Simplifying Radical Expressions Involving Fractions

1. *Division of Radicals* ☞ $\sqrt[n]{\frac{a}{b}} = \frac{\sqrt[n]{a}}{\sqrt[n]{b}}$ where a, b are non-negative numbers and $b \neq 0$.

Multiplying Radical Expressions

1. *Product of Radicals* ☞ $k_1\sqrt[n]{a} \times k_2\sqrt[n]{b} = k_1k_2\sqrt[n]{ab}$, where $a, b \geq 0$.

Adding and Subtracting Radical Expressions

1. *Simplifying Like Radicals* ☞ $a\sqrt[n]{b} + c\sqrt[n]{b} = (a+c)\sqrt[n]{b}$ where $b \geq 0$.

Domain and Range of Radical Functions

1. *Domain of a Radical Function* For $y = c\sqrt{f(x)} + k$, the domain includes all x such that $f(x) \geq 0$.
2. *Range of a Radical Function* For $y = c\sqrt{f(x)} + k$, the range is: $y \geq k$ if $c > 0$ and $y \leq k$ if $c < 0$.

Formulas For Chapter Rational and Irrational Expressions

Rational and Irrational Numbers

1. *Rational Numbers Definition* A number expressible as $\frac{p}{q}$ with $p, q \in \mathbb{Z}$ and $q \neq 0$.
2. *Irrational Numbers Definition* A real number that cannot be written as $\frac{p}{q}$ with $p, q \in \mathbb{Z}$ and $q \neq 0$.

Multiplying Rational Expressions

1. *Multiplication of Rational Expressions* $\frac{a}{b} \times \frac{c}{d} = \frac{a \times c}{b \times d}$, then factor, cancel common factors and simplify.

Dividing Rational Expressions

1. *Division of Rational Expressions* $\frac{a}{b} \div \frac{c}{d} = \frac{a}{b} \times \frac{d}{c} = \frac{a \times d}{b \times c}$, then factor, cancel common factors and simplify.

Rational Equations

1. *Common Denominator Method* $\frac{a}{b} = \frac{c}{d} \Rightarrow \frac{ad}{bd} = \frac{cb}{bd}$
2. *Cross-Multiplication* $\frac{a}{b} = \frac{c}{d} \Rightarrow ad = bc$

Simplifying Complex Fractions

1. *Converting to Improper Fractions* Mixed number $a\frac{b}{c}$ to improper fraction $\frac{ac+b}{c}$.
2. *Division of Fractions (Keep, Change, Flip)* $\frac{a}{b} \div \frac{c}{d} = \frac{a}{b} \times \frac{d}{c} = \frac{ad}{bc}$.

Solving Rational Inequalities

1. *General Form of a Rational Inequality* $\frac{P(x)}{Q(x)} < 0$ or $\frac{P(x)}{Q(x)} \leq 0$ or $\frac{P(x)}{Q(x)} > 0$ or $\frac{P(x)}{Q(x)} \geq 0$ where $P(x)$ and $Q(x)$ are polynomials, and $Q(x) \neq 0$.
2. *Identifying Critical Points* Solve $P(x) = 0$ and $Q(x) = 0$ for x to find the critical points.

Irrational Functions

1. *General Form of an Irrational Function* $f(x) = \sqrt[n]{(g(x))^m}$ or $f(x) = (g(x))^{\frac{m}{n}}$
2. *Domain of an Irrational Function (Odd Index)* For odd n, the domain of $f(x)$ is the same as the domain of $g(x)$.
3. *Domain of an Irrational Function (Even Index)* For even n and any integer m, the domain of $f(x)$ is: $\{x \in \text{Domain of } g \mid (g(x))^m \geq 0\}$.

Direct, Inverse, Joint, and Combined Variation

1. *Direct Variation* $y = kx$ where k is the constant of variation.
2. *Inverse Variation* $y = \frac{k}{x}$ where k is the constant of variation.
3. *Joint Variation* $y = kxz$ where k is the constant of variation.
4. *Combined Variation* $y = k\left(\frac{x}{z}\right)$ where k is the constant of variation.

Formulas For Chapter Trigonometric Functions

Angles of Rotation

1. *Finding Coterminal Angles* $\theta' = \theta \pm 360°k$ or $\theta' = \theta \pm 2k\pi$ for any integer k

Angles and Angle Measure

1. *Converting Degrees to Radians* ☞ $\text{Radian} = \text{Degree} \times \frac{\pi}{180}$
2. *Converting Radians to Degrees* ☞ $\text{Degree} = \text{Radian} \times \frac{180}{\pi}$

Right-Triangle Trigonometry

1. *Pythagorean Theorem* ☞ $c^2 = a^2 + b^2$ where c is the hypotenuse and a, b are the legs of the right triangle.

Trigonometric Ratios

1. *Sine of an angle* ☞ $\sin(\theta) = \frac{\text{opposite}}{\text{hypotenuse}}$
2. *Cosine of an angle* ☞ $\cos(\theta) = \frac{\text{adjacent}}{\text{hypotenuse}}$
3. *Tangent of an angle* ☞ $\tan(\theta) = \frac{\text{opposite}}{\text{adjacent}}$

Function Values and Ratios of Special and General Angles

1. *Values at 0 Degrees* ☞ $\sin(0^\circ) = 0$, $\cos(0^\circ) = 1$, $\tan(0^\circ) = 0$
2. *Values at 30 Degrees* ☞ $\sin(30^\circ) = \frac{1}{2}$, $\cos(30^\circ) = \frac{\sqrt{3}}{2}$, $\tan(30^\circ) = \frac{1}{\sqrt{3}}$
3. *Values at 45 Degrees* ☞ $\sin(45^\circ) = \frac{\sqrt{2}}{2}$, $\cos(45^\circ) = \frac{\sqrt{2}}{2}$, $\tan(45^\circ) = 1$
4. *Values at 60 Degrees* ☞ $\sin(60^\circ) = \frac{\sqrt{3}}{2}$, $\cos(60^\circ) = \frac{1}{2}$, $\tan(60^\circ) = \sqrt{3}$
5. *Values at 90 Degrees* ☞ $\sin(90^\circ) = 1$, $\cos(90^\circ) = 0$, $\tan(90^\circ) =$ "undefined"
6. *Complementary Angle Identities* ☞ $\sin(\theta) = \cos(90^\circ - \theta)$ and $\cos(\theta) = \sin(90^\circ - \theta)$
7. *Even-Odd Identities* ☞ $\cos(-\theta) = \cos(\theta)$ and $\sin(-\theta) = -\sin(\theta)$

The Reciprocal Trigonometric Functions

1. *Secant Function* ☞ $\sec(\theta) = \frac{1}{\cos(\theta)}$, for $\cos(\theta) \neq 0$.
2. *Cosecant Function* ☞ $\csc(\theta) = \frac{1}{\sin(\theta)}$, for $\sin(\theta) \neq 0$.
3. *Cotangent Function* ☞ $\cot(\theta) = \frac{1}{\tan(\theta)}$, for $\tan(\theta) \neq 0$.

Co-functions

1. *Sine Co-function Identity* ☞ $\sin(90° - x) = \cos(x)$
2. *Cosine Co-function Identity* ☞ $\cos(90° - x) = \sin(x)$
3. *Tangent Co-function Identity* ☞ $\tan(90° - x) = \cot(x)$
4. *Cotangent Co-function Identity* ☞ $\cot(90° - x) = \tan(x)$
5. *Secant Co-function Identity* ☞ $\sec(90° - x) = \csc(x)$
6. *Cosecant Co-function Identity* ☞ $\csc(90° - x) = \sec(x)$

Reference Angles and the Calculator

1. *Quadrant Function Signs* ☞ First Quadrant: All functions positive. Second Quadrant: sin positive, cos and tan negative. Third Quadrant: sin and cos negative, tan positive. Fourth Quadrant: cos positive, sin and tan negative.
2. *Finding Reference Angles* ☞ First Quadrant: θ. Second Quadrant: $180° - \theta$. Third Quadrant: $\theta - 180°$. Fourth Quadrant: $360° - \theta$.

Pythagorean Identities

1. *Fundamental Pythagorean Identity* ☞ $\sin^2\theta + \cos^2\theta = 1$ for any angle θ.
2. *Derived Pythagorean Identity (by dividing by $\sin^2\theta$)* ☞ $1 + \cot^2\theta = \csc^2\theta$
3. *Derived Pythagorean Identity (by dividing by $\cos^2\theta$)* ☞ $\tan^2\theta + 1 = \sec^2\theta$

The Unit Circle, Sine, and Cosine

1. *Unit Circle Equation* ☞ $x^2 + y^2 = 1$
2. *Sine from Unit Circle* ☞ $\sin\theta = y$-coordinate of point on the unit circle
3. *Cosine from Unit Circle* ☞ $\cos\theta = x$-coordinate of point on the unit circle

Arc Length and Sector Area

1. *Area of a Sector* ☞ $A = \pi r^2 \left(\frac{\theta}{360}\right)$ where r is the radius, and θ is the central angle in degrees.
2. *Arc Length* ☞ $L = \pi r \left(\frac{\theta}{180}\right)$ where r is the radius, and θ is the central angle in degrees.

Formulas For Chapter Sequences and Series

Arithmetic Sequences

1 *n-th Term of an Arithmetic Sequence* $x_n = a + (n-1)d$ where a is the first term, d is the common difference, and n is the term number.

Geometric Sequences

1 *General Form of a Geometric Sequence* $x_n = ar^{(n-1)}$ where x_n is the n-th term, a is the first term, and r is the common ratio.

Finite Arithmetic Series

1 *Sum of a Finite Arithmetic Series* $S_n = \frac{1}{2}n(2a + d(n-1))$ where n is the number of terms, a is the first term, and d is the common difference.

2 *Concise Sum of a Finite Arithmetic Series* $S_n = \frac{1}{2}n(a + x_n)$ where n is the number of terms, a is the first term, and x_n is the n-th term.

Finite Geometric Series

1 *Sum of a Finite Geometric Series* $S_n = a\left(\frac{1-r^n}{1-r}\right)$ where n is the number of terms, a is the first term, and r is the common ratio.

Infinite Geometric Series

1 *Sum of an Infinite Geometric Series* $S = \frac{a}{1-r}$ where a is the first term and r is the common ratio $(|r| < 1)$.

Pascal's Triangle

1 *Pascal's Triangle nth Entry* The k-th entry of the n-th row is $\binom{n}{k} = \frac{n!}{k!(n-k)!}$.

Binomial Theorem

1. *Binomial Theorem Formula* $(x+y)^n = \sum_{k=0}^{n} \binom{n}{k} x^{n-k} y^k$
2. *Binomial Coefficient* $\binom{n}{k} = \dfrac{n!}{k!(n-k)!}$
3. *k-th Term of a Binomial Expansion* The k-th term is given by $\binom{n}{k-1} x^{n-k+1} y^{k-1}$

Sigma Notation (Summation Notation)

1. *Definition of Sigma Notation* $\sum_{i=1}^{n} f(i) = f(1) + f(2) + \cdots + f(n)$ where i an indexing variable and $f(i)$ is a function of i.

Alternate Series

1. *General Form of an Alternating Series* $\sum_{k=1}^{\infty} (-1)^k a_k$ where $a_k \geq 0$.
2. *Convergence of an Alternating Series* An alternating series converges if $0 \leq a_{n+1} \leq a_n$ for all $n \geq 1$ and $a_n \to 0$ as $n \to \infty$.

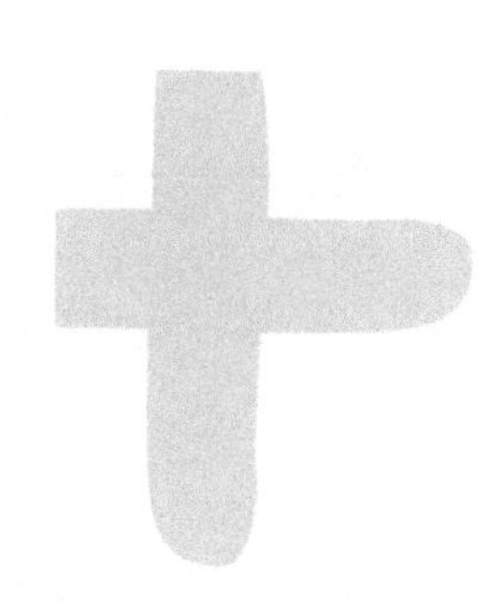

15. Practice Test 1

15.1 Practices

1) Which of the following is a factor of $12x^8 - 21x^4 + 3x^3$?

☐ A. $4x^5 - 7x + 1$

☐ B. $x - 1$

☐ C. $3x^2 - x - 1$

☐ D. $3x + 1$

☐ E. $4x^5 - 7x^2$

2) If $f(x) = 3x + 4(x+1) + 2$, then $f(4x) = ?$

☐ A. $28x + 6$

☐ B. $16x - 6$

☐ C. $25x + 4$

☐ D. $12x + 3$

☐ E. $12x + 28$

3) Simplify $(-4+9i)(3+5i)$.

☐ A. $-57 - 7i$

☐ B. $-54 + 7i$

☐ C. $57 - 7i$

☐ D. $-57 + 7i$

☐ E. $57+54i$

4) Simplify this expression: $\frac{(-2x^2y^2)^3(3x^3y)}{12x^3y^8}$.

☐ A. $\frac{-6x^6y^7}{2y}$

☐ B. $\frac{-3x^6}{2y}$

☐ C. $\frac{-2x^6}{y}$

☐ D. $\frac{12x^6}{y^8}$

☐ E. $\frac{-x^6y^7}{y}$

5) Solve and write the solution set in set-builder notation of the following inequality.

$$2x-4(x+2)\geq x+6.$$

☐ A. $\{x|x\leq\frac{2}{3}\}$

☐ B. $\{x|x\geq\frac{2}{3}\}$

☐ C. $\{x|x\geq\frac{14}{3}\}$

☐ D. $\{x|x\geq\frac{-14}{3}\}$

☐ E. $\{x|x\leq\frac{-14}{3}\}$

6) Which of the following equations is equivalent to $3^{2x}=5$?

☐ A. $x=\frac{\log_3 5}{2}$

☐ B. $x=\frac{\log_5 2}{3}$

☐ C. $x=\frac{\log_2 3}{5}$

☐ D. $x=\log_3\left(\frac{5}{2}\right)$

☐ E. $x=\frac{\log_5 3}{2}$

7) Which equation represents the solution for x in the formula $3(7^{2x})=11$?

☐ A. $x=\frac{1}{2}\left(\frac{\log 11}{\log 3}-\log 7\right)$

☐ B. $x=\frac{2(\log 11-\log 3)}{\log 7}$

☐ C. $x=2\left(\frac{\log 11}{\log 3}-\log 7\right)$

☐ D. $x=\frac{\log 11-\log 3}{2\log 7}$

☐ E. $x=\frac{2(\log 3-\log 7)}{\log 11}$

8) What is the negative solution to $2x^3 - x^2 - 6x = 0$? Write your answer in the box. ☐

9) If $y = 3x^2 + 6x + 2$ is graphed in the xy-plane, which of the following characteristics of the graph is displayed as a constant or coefficient in the equation?

☐ A. y-coordinate of the vertex

☐ B. x-intercept(s)

☐ C. y-intercept

☐ D. x-intercept of the symmetry

☐ E. x-coordinate of the vertex

10) The given expression $2x - 3$ is a factor of which equation?

☐ A. $(2x - 3) + 17$

☐ B. $2x^3 - x^2 - 3x$

☐ C. $6(3 + 2x)$

☐ D. $3x^2 - 4x + 7$

☐ E. $3x^3 - 4x + 3$

11) The table below illustrates the linear relationship between the number of gadgets sold at an electronics store and the revenue generated in thousand dollars.

Number of Gadgets Sold (per hundred)	2	5	9	12	14
Revenue Generated (in Thousand Dollars)	10	22	38	50	58

Determine the rate of change in revenue generated in thousand dollars with respect to the number of gadgets sold in the store.

☐ A. 3

☐ B. 4

☐ C. 7

☐ D. 12

☐ E. 14

12) What is the zero of $p(x) = \frac{5}{9}x + 15$?

- ☐ A. -27
- ☐ B. -15
- ☐ C. 15
- ☐ D. 27
- ☐ E. 30

13) A member pays an annual membership fee of \$150 to a book club. Each time they purchase a book, they pay an additional \$8. The total amount of money they spend at the book club in a year can be calculated using the function $g(x) = 8x + 150$. What does the variable x represent in this function?

- ☐ A. The total amount of money the member spends on books each month.
- ☐ B. The number of months the member has been in the book club.
- ☐ C. The initial membership fee paid by the member.
- ☐ D. The number of books the member purchases in one year.
- ☐ E. The cost of each book purchased by the member.

14) Which expression is equivalent to $16m^2 - 64$?

- ☐ A. $(4m-8)(m-4)$
- ☐ B. $16(m-2)(m-2)$
- ☐ C. $16m(m-4)$
- ☐ D. $16(m-2)(m+2)$
- ☐ E. $16(m-4)(m+4)$

15) How was the graph of $f(x) = \pi^x$ transformed to create the graph of $h(x) = \pi^{x-2}$?

- ☐ A. A horizontal shift right and a vertical shift down
- ☐ B. A vertical shift up
- ☐ C. A horizontal shift to the right
- ☐ D. A vertical shift down
- ☐ E. A horizontal shift to the left

16) The sum of $-5x^2 + 4x - 30$ and $9x^2 - 6x + 20$ can be written in the form $ax^2 + bx + c$, where a, b, and c are constants. What is the value of $a + b - c$? Write your answer in the box. ☐

17) The graph of the parent function $g(x) = x^2$ was transformed to create the graph of $m(x) = g(x-3)+2$. Which graph best represents m?

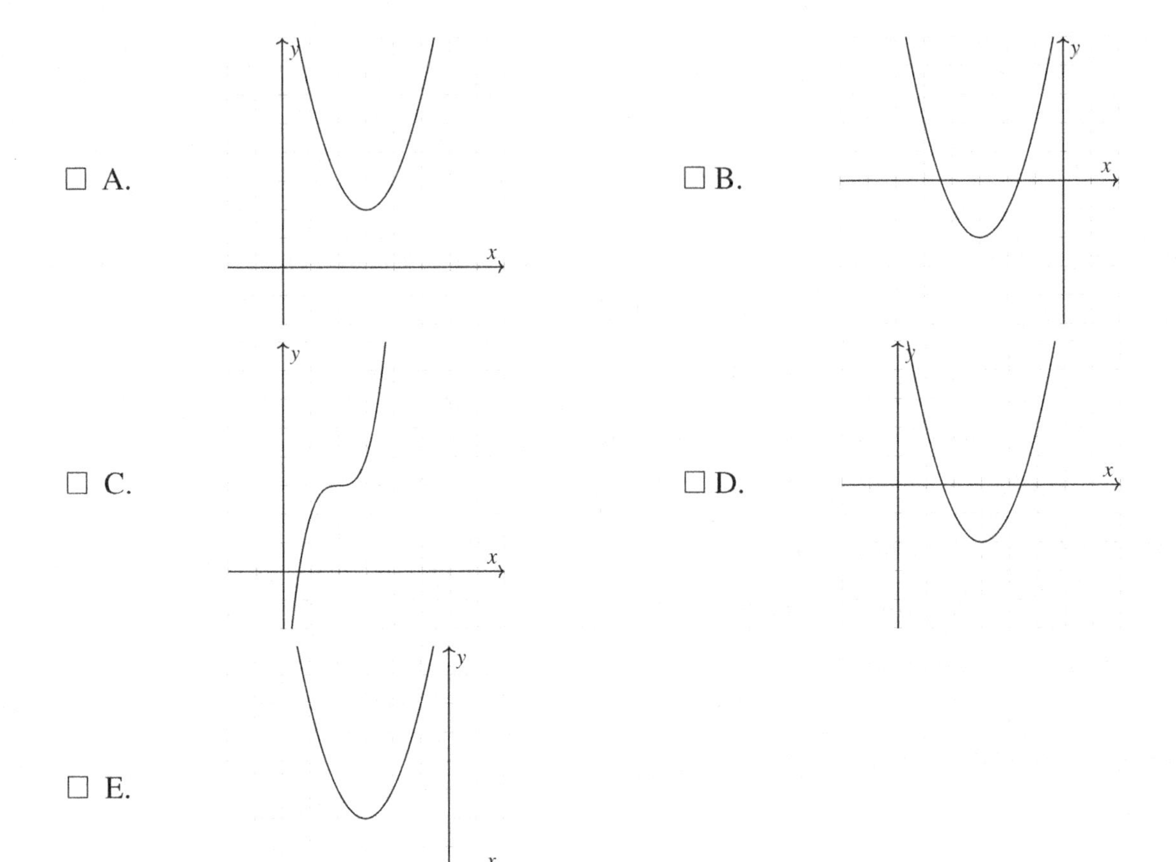

18) What is the vertex of the graph of the quadratic function $h(x) = 2x^2 + 6x + 3$?

☐ A. $\left(-\frac{6}{2}, \frac{3}{2}\right)$

☐ B. $\left(-\frac{6}{2}, \frac{6}{2}\right)$

☐ C. $\left(-\frac{3}{2}, \frac{3}{2}\right)$

☐ D. $\left(-\frac{3}{2}, -\frac{3}{2}\right)$

☐ E. $\left(\frac{6}{2}, -\frac{6}{2}\right)$

19) An investment account grows according to the function $f(x) = 5000(1.08)^x$, where x is the number of years. Which statement is the best interpretation of the value 1.08 in this function?

☐ A. The account shrinks by 8% per year.

☐ B. The account shrinks at a rate of 8% per year.

☐ C. The account shrinks by 92% per year.

☐ D. The account grows by $1.08 per year.

☐ E. The account grows at a rate of 8% per year.

20) What are the zeros of the function: $f(x) = x^2 - 7x + 12$?

☐ A. $-4, -3$

☐ B. $-2, -3$

☐ C. 0

☐ D. 4, 3

☐ E. 0, 3, 4

21) Find the factors of the binomial $x^3 - 8$.

☐ A. $(x+2)(x-2)$

☐ B. $(x^2+4)(x^2-4)$

☐ C. $(x^2+2x+4)(x-2)$

☐ D. $(x^2+2x+4)(x+2)$

☐ E. $(x^2-4)(x-4)$

22) The measure of an acute angle is represented by $(5x-20)^{\circ}$. Which value is not a possible value for x?

☐ A. 10.6

☐ B. 20.8

☐ C. 23

☐ D. 21

☐ E. 18

23) What is the value of the y-intercept of the graph of $k(x) = 42\left(\frac{4}{5}\right)^x$? Write your answer in the box.

☐

24) A musician has two gigs this month. The total time dedicated to both gigs cannot exceed 150 hours for the month. Which graph best represents the solution set for all possible combinations of x, the number of hours dedicated to the first gig, and y, the number of hours dedicated to the second gig, in one month?

☐ A.

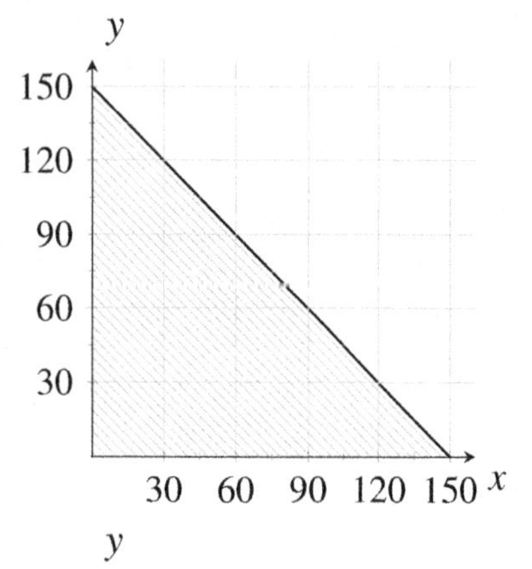

☐ B.

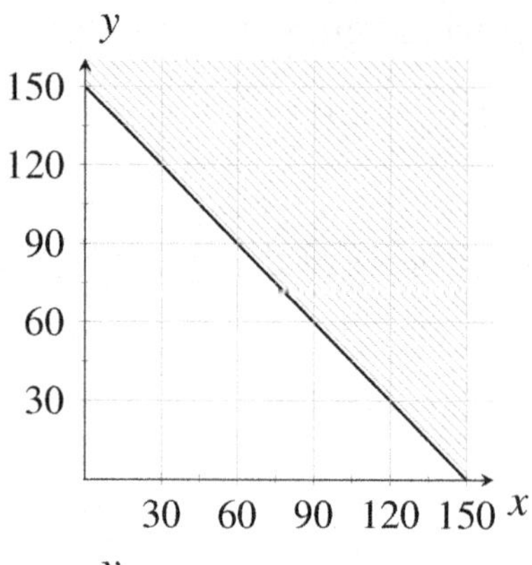

☐ C.

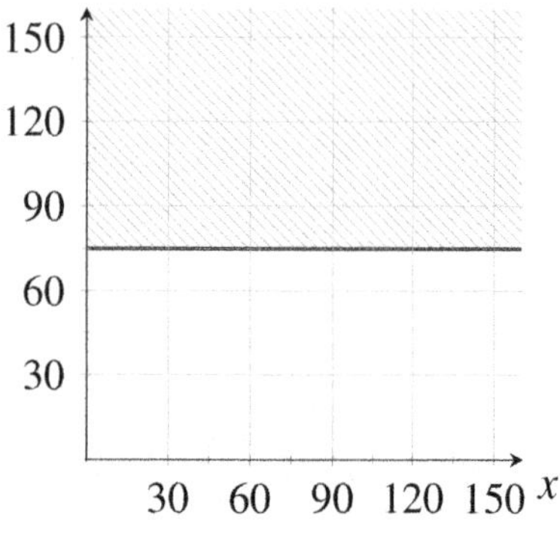

☐ D.

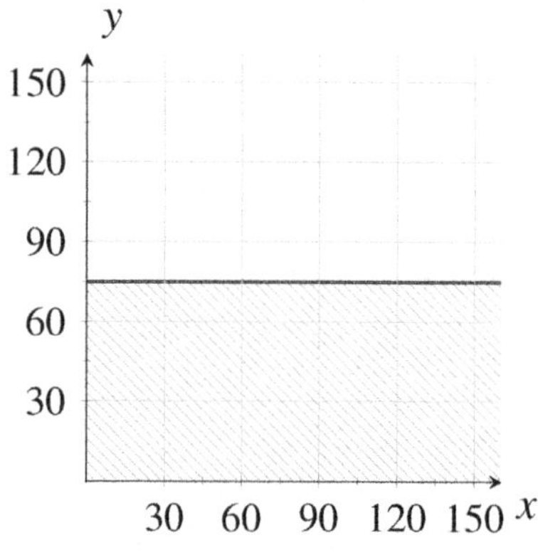

☐ E.

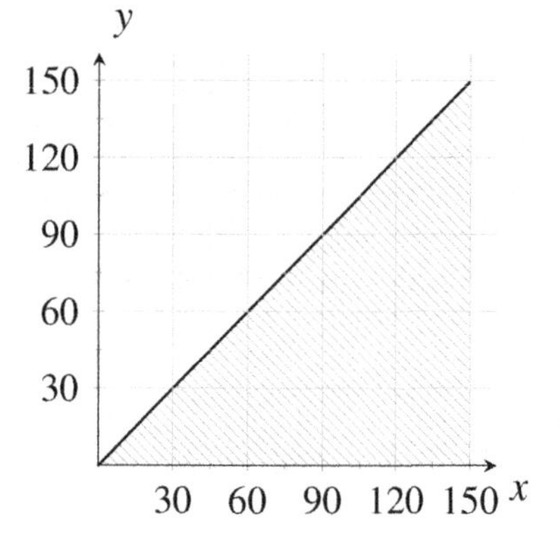

25) Which of the following is equivalent to $(3n^2+2n+6)-(2n^2-4)$?

☐ A. $n-2$

☐ B. n^2-3

☐ C. n^2+2n-2

☐ D. $n^2+2n+10$

☐ E. $5n^2+2n+1$

26) The population of buffaloes in the 18th century is modeled by an exponential function f, where x is the number of decades after the year 1850. The graph of f is shown on the grid.

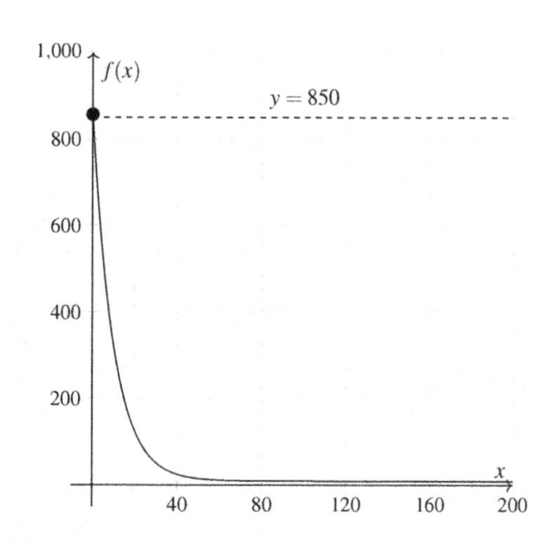

Which inequality best represents the range of f in this situation?

☐ A. $x > 0$

☐ B. $y \leq 850$

☐ C. $0 \leq x \leq 1{,}960$

☐ D. $0 < y \leq 850$

☐ E. $0 \leq y \leq 1{,}960$

27) Which ordered pair is in the solution set of $3y - 4x \geq 6$?

☐ A. $(2,2)$

☐ B. $(0,2)$

☐ C. $(2,0)$

☐ D. $(3,1)$

☐ E. $(-1,-2)$

28) If $2^p.4^q = 2^{10}$, what is the value of $p + 2q$?

☐ A. 10

☐ B. 8

☐ C. 6

☐ D. 5

☐ E. 4

29) Which system of equations is best represented by lines h and i?

Line h

x	0	1	0.5
y	−0.5	0.5	0

Line i

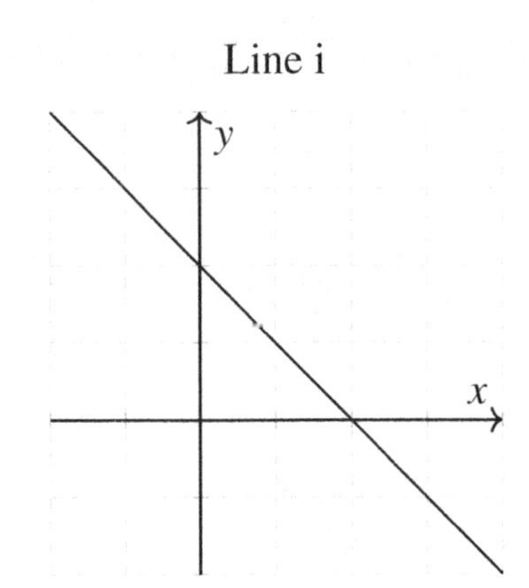

☐ A. $\begin{cases} y = \frac{1}{2}x + \frac{1}{2} \\ y = 2 - x \end{cases}$

☐ B. $\begin{cases} 2x - 2y = 1 \\ x + y = 2 \end{cases}$

☐ C. $\begin{cases} y = \frac{2}{3}x - 2 \\ y = \frac{1}{2} - x \end{cases}$

☐ D. $\begin{cases} 2x - 2y = 1 \\ x - y = 2 \end{cases}$

☐ E. $\begin{cases} y = \frac{2}{3}x - 1 \\ y = \frac{1}{2} + x \end{cases}$

30) A circle is inscribed in a square and the radius of the circle is 4. What is the area of the shaded region?

☐ A. $16 - 64\pi$

☐ B. $16 + 16\pi$

☐ C. $64 - 16\pi$

☐ D. $16 - 16\pi$

☐ E. 4π

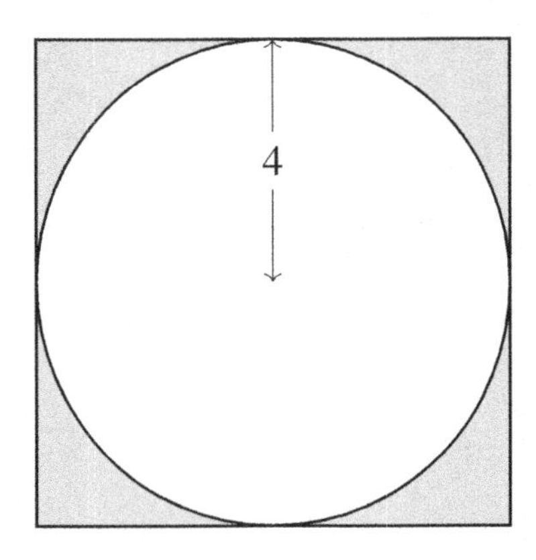

31) In an archery school, the coach has 107 arrows for the players. The coach will give each player on the team 9 arrows. The graph shows the linear relationship between y, the number of arrows of players, and x, the number of players.

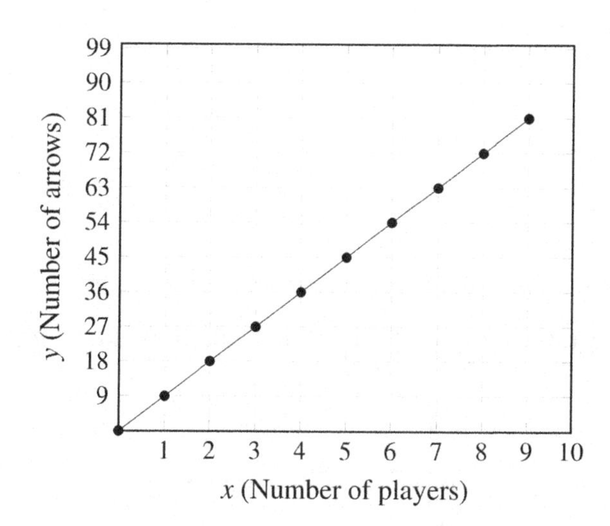

The coach will use no more than 9 players for the team. Which set best represents the domain of the function for this situation?

☐ A. $\{9,18,27,36,45,54,63,72,81\}$

☐ B. $\{107,98,89,80,71,62,53,44,35\}$

☐ C. $\{0,1,2,3,4,5,6,7,8,9\}$

☐ D. $\{1,2,3,4,5,6,7,8,9,10\}$

☐ E. $\{0,81\}$

32) Given $q(x)=3(x-5)^2-7$, what is the value of $q(2)$? Write your answer in the box. ☐

33) The table shows the value in thousands of dollars of a car at the end of x years.

Number of Years, x	1	2	3	4	5
Value, v(x), (thousand dollars)	129	141	150	169	188

Which exponential function models this situation?

☐ A. $v(x)=161(1.1)^x$

☐ B. $v(x)=115(1.1)^x$

☐ C. $v(x)=1.1(0.95)^x$

☐ D. $v(x)=85(0.9)^x$

☐ E. $v(x)=90(0.95)^x$

34) Expand the expression $(7x+2y)(5x+2y)$.

☐ A. $2x^2+4xy+2y^2$

☐ B. $12x^2+14xy+4y$

☐ C. $7x^2+14xy+y^2$

☐ D. $32x^2+24xy+y$

☐ E. $35x^2+24xy+4y^2$

35) An exponential function is graphed on the grid. Which function is best represented by the graph?

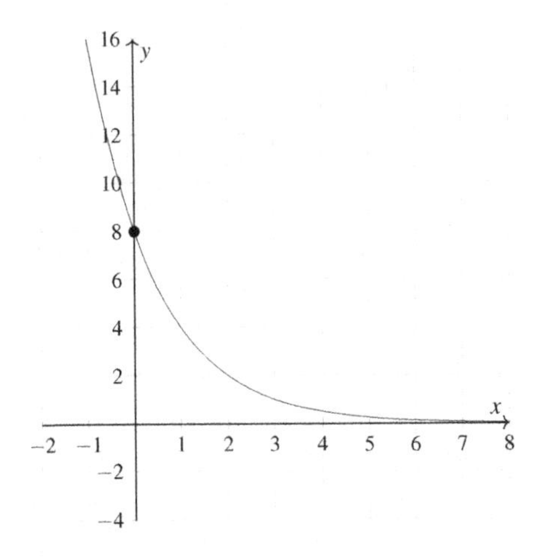

☐ A. $g(x)=8-(2)^x$

☐ B. $g(x)=4\left(\frac{1}{2}\right)^x$

☐ C. $g(x)=8\left(\frac{1}{2}\right)^x$

☐ D. $g(x)=4(2)^x$

☐ E. $g(x)=8(2)^x$

36) In the xy-plane, if $(0,0)$ is a solution to the system of inequalities below, which of the following relationships between a and b must be true?

$$\begin{cases} y<x+a \\ y>x+2b \end{cases}$$

☐ A. $b>a$

☐ B. $a=2b$

☐ C. $|a|>|b|$

☐ D. $a=b$

☐ E. $a>b$

37) Multiply and write the product in scientific notation:

$$(2.9 \times 10^{6}) \times (2.6 \times 10^{-5}).$$

☐ A. 75.4×10^{-5}

☐ B. 0.754

☐ C. 7.54×10^{1}

☐ D. 754×100

☐ E. 75.4×10^{11}

38) Which of the following is equivalent to $-2 + \frac{3b-4c}{9b} - \frac{2b+2c}{6b}$?

☐ A. $\frac{-18b-7c}{9b}$

☐ B. $\frac{-18b-7c}{18b}$

☐ C. $\frac{-18b-14c}{18b}$

☐ D. $\frac{18b-14c}{18b}$

☐ E. $\frac{18b-7c}{9b}$

39) Which graph represents $y = 2x^2 - 4x - 1$?

☐ A.

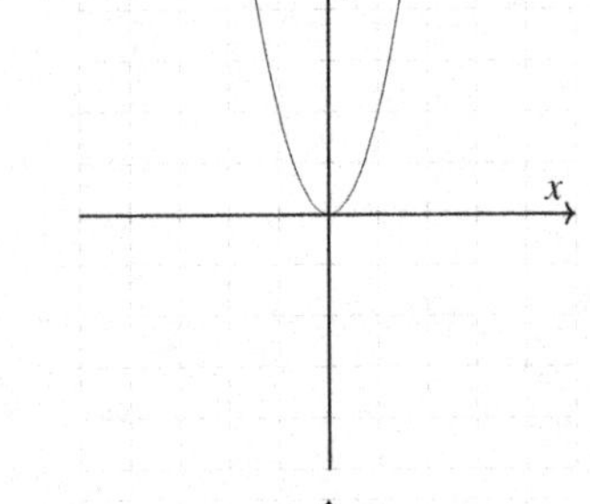

☐ B.

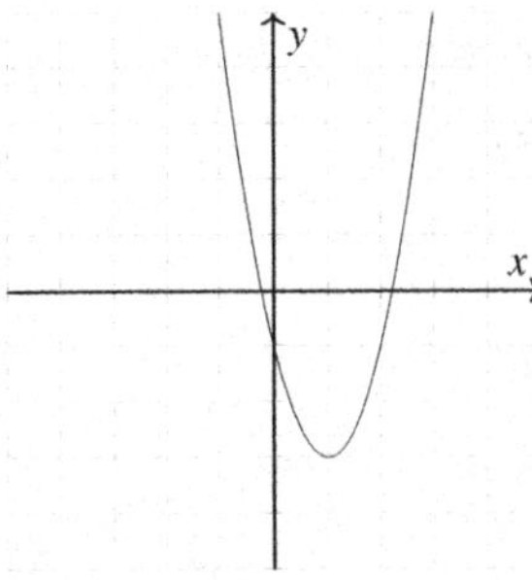

☐ C.

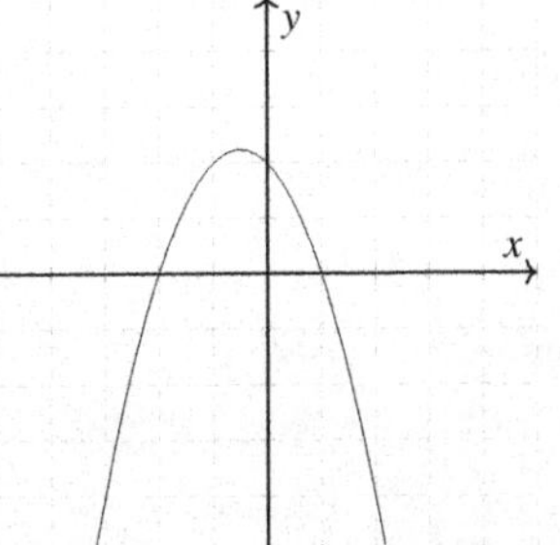

☐ D.

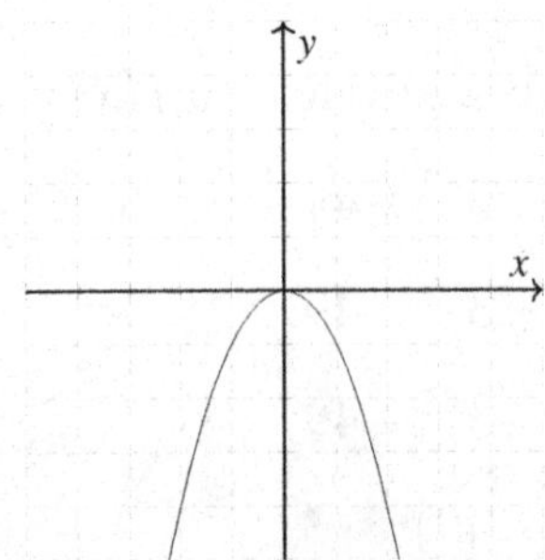

☐ E.

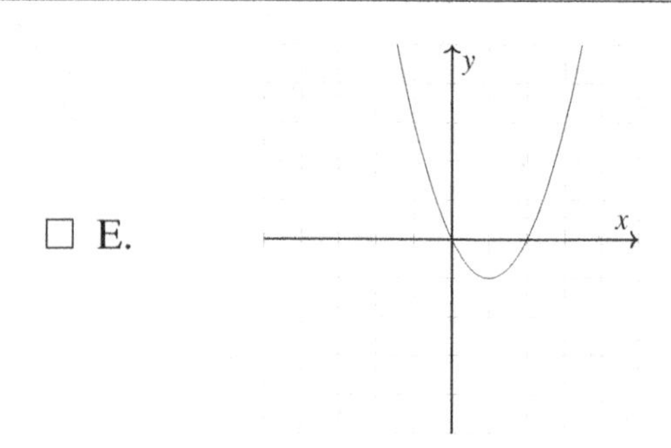

40) What are the solutions to the equation $x^2 - 3x + 1 = x - 3$? Write your answer in the box.

41) Multiply and write the product in scientific notation:

$$(1.7 \times 10^9) \times (2.3 \times 10^{-7})$$

☐ A. 3.91×10^{-2}

☐ B. 3.91×10

☐ C. 0.391×10^2

☐ D. 3.91×10^2

☐ E. 39.1×10^2

42) The equation $h = -25t^2 + st + k$ gives the height h, in feet, of a ball t seconds after it is thrown straight up with an initial speed of s feet per second from a height of k feet. Which of the following gives s in terms of h, t, and k?

☐ A. $s = h + k + 25t$

☐ B. $s = \frac{h-k}{t} + 25t$

☐ C. $s = \frac{h-k+25}{t}$

☐ D. $s = \frac{h-k}{t} - 25t$

☐ E. $s = h + k - 25t$

43) Which function is equivalent to $f(x) = 5x^2 - 30x - 4$?

☐ A. $f(x) = 5(x-3)^2 + 49$

☐ B. $f(x) = 5(x+3)^2 - 49$

☐ C. $f(x) = 5(x-3)^2 - 49$

☐ D. $f(x) = 5(x-49)^2 - 3$

☐ E. $f(x) = 5(x-49)^2 + 3$

44) The graph shows the height in feet of a ball above the ground t seconds after it was launched from a height of 50 feet. Which function is best represented by the graph of this situation?

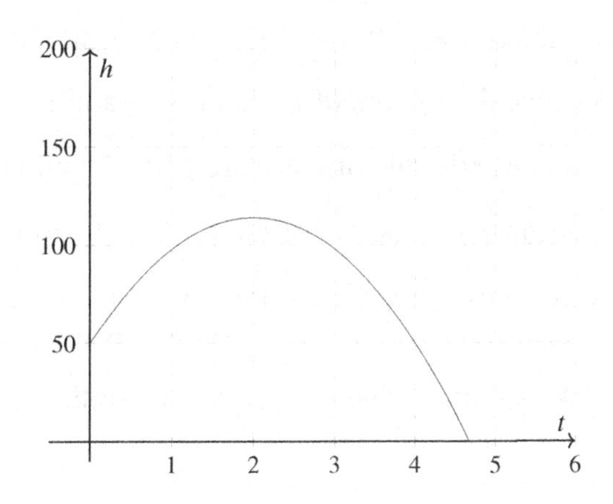

☐ A. $y = -16t^2 + 64t + 50$

☐ B. $y = -16t^2 - 64t + 50$

☐ C. $y = -16t^2 + 128t - 100$

☐ D. $y = 16t^2 - 128t - 100$

☐ E. $y = 16t^2 - 100$

45) Which of the following expressions is equivalent to $2x(5 + 3y + 2x + 4z)$?

☐ A. $7x + 5xy + 4x + 6xz$

☐ B. $5x + 3xy + 4x^2 + 4xz$

☐ C. $6xy + 4x^2 + 8xz + 10$

☐ D. $10x + 6xy + 4x^2 + 8xz$

☐ E. $7x + 5xy + 4x^2 + 6xz + 10$

46) What is the value of x in the following equation?

$$\log_4(x+2) - \log_4(x-2) = 1.$$

☐ A. 10

☐ B. $\log_4 10$

☐ C. $\log_{10} 4$

☐ D. $\frac{10}{3}$

☐ E. $-\frac{10}{3}$

47) Given $g(x) = 3x^2 + 20x - 7$, which statement is true?

☐ A. The zeroes are 3 and -7, because the factors of g are $(x-3)$ and $(x+7)$.

☐ B. The zeroes are 3 and 7, because the factors of g are $(x-3)$ and $(x-7)$.

☐ C. The zeroes are $\frac{1}{3}$ and -7, because the factors of g are $(3x+1)$ and $(x-7)$.

☐ D. The zeroes are $\frac{1}{3}$ and -7, because the factors of g are $(3x-1)$ and $(x+7)$.

☐ E. The zeroes are $\frac{1}{3}$ and 7, because the factors of g are $(3x+1)$ and $(x+7)$.

48) A customer purchased movie tickets online. The total cost, c, in dollars, of t tickets can be found using the function below:

$$c = 24.50t + 5.25.$$

If the customer spent a total of \$103.25 on tickets, how many tickets did he purchase?

☐ A. 4

☐ B. 5

☐ C. 6

☐ D. 7

☐ E. 8

49) Mrs. Johnson has a container holding m liters of a liquid for her biology class experiment. If she allocates 2 liters to each student, she has 4 liters remaining. To give 3 liters to each student, she needs an additional 18 liters. How many students are in her class?

☐ A. 10

☐ B. 14

☐ C. 18

☐ D. 20

☐ E. 22

50) Determine the y-intercepts and x-intercept for the equation $2x + 4y = 16$.

☐ A. $(0,4)$, $(8,0)$

☐ B. $(0,4)$, $(-8,0)$

☐ C. (0,8), (4,0)

☐ D. (0,−4), (8,0)

☐ E. (0,−4), (−8,0)

51) Estimate the perimeter of the shape illustrated below, where $\pi \approx 3$.

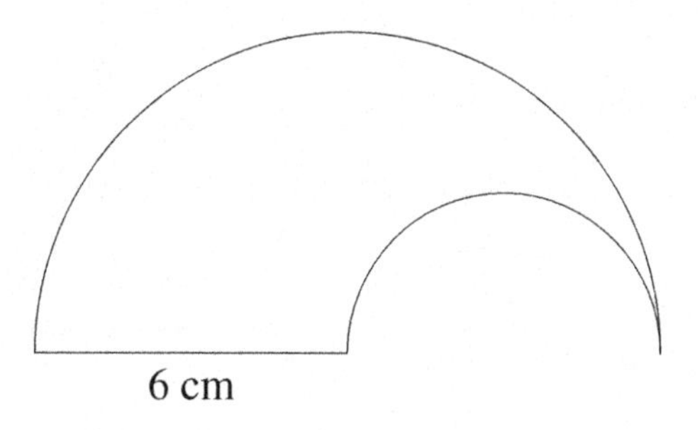

☐ A. $27cm$

☐ B. $31cm$

☐ C. $33cm$

☐ D. $36cm$

☐ E. $39cm$

52) The graph below represents a quadratic function. Determine the equation that correctly represents the axis of symmetry for this function.

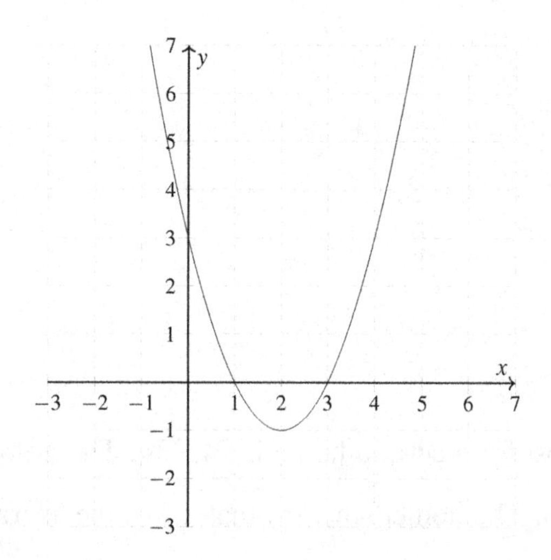

☐ A. $y = 2x + 1$

☐ B. $x = 2y - 1$

☐ C. $y = 2$

☐ D. $x = 2$

☐ E. $x = 0$

53) What is the inverse of the function $f(x) = x^2 + 1$?

☐ A. $f^{-1}(x) = \sqrt{x} - 1$

☐ B. $f^{-1}(x) = \frac{1}{\sqrt{x-1}}$

☐ C. $f^{-1}(x) = \frac{1}{x^2-1}$

☐ D. $f^{-1}(x) = \pm\sqrt{x-1}$

☐ E. $f^{-1}(x) = \pm\frac{1}{\sqrt{x-1}}$

54) An engineer is designing a suspension bridge with n sections. The length of each section is 48.2 m. The bridge will also have two towers, each of which is 125.6 m tall. Which function can be used to find the total length of the bridge in meters, including the tower?

☐ A. $L(n) = 251.2 + 48.2n$

☐ B. $L(n) = 48.2n + 125.6$

☐ C. $L(n) = 251.2n$

☐ D. $L(n) = 173.8n$

☐ E. $L(n) = 48.2n - 125.6$

55) Perform the operations and simplify $\sqrt{8} - \sqrt{50} + \sqrt{72}$.

☐ A. $13\sqrt{2}$

☐ B. $6\sqrt{2}$

☐ C. $3\sqrt{2}$

☐ D. $\sqrt{2}$

☐ E. $-6\sqrt{2}$

56) The average annual energy cost for a certain home is $4,334. The homeowner plans to spend $25,000 to install a geothermal heating system. The homeowner estimates that the average annual energy cost will then be $2,712. Which of the following inequalities can be solved to find t, the number of years after installation at which the total amount of energy cost saving will exceed the installation cost?

☐ A. $25,000 > (4,334 - 2,712)t$

☐ B. $25,000 < (4,334 - 2,712)t$

☐ C. $25,000 - 4,334 < 2,712t$

☐ D. $25,000 > 4,332 - 2.712t$

☐ E. $25,000 + 4,334 > 2,712t$

57) The equation $y = x^2 - 9x + 18$ represents a parabola in the xy-plane. Determine the x-intercepts of the parabola. Write your answer in the box. ☐

58) What is the solution to the system of equations below?

$$\begin{cases} 3x - 5y = 8 \\ 9x - 15y = 24 \end{cases}$$

☐ A. The ordered pair $\left(\frac{8}{3}, 0\right)$ is the solution.

☐ B. The ordered pair $\left(0, \frac{8}{5}\right)$ is the solution.

☐ C. The ordered pair $(2,4)$ is the solution.

☐ D. There is no solution.

☐ E. There are an infinite number of solutions.

59) Which statement about the quadratic functions below is true?

$$f(x) = 2x^2 - 4, \quad g(x) = -x^2 + 6 \quad \text{and} \quad h(x) = x^2 + x - 1.$$

☐ A. The graphs of two of these functions have the same maximum point.

☐ B. The graphs of two of these functions have the same axis of symmetry.

☐ C. The graphs of two of these functions intersect each other on the x-axis.

☐ D. The graphs of all these functions have the same y-intercepts.

☐ E. The functions do not intersect with each other.

60) What is the domain of $f(x) = -3x^2 + 16$?

☐ A. $(-\infty, 16]$

☐ B. $(-4, 4)$

☐ C. $\left[-\frac{4}{3}, \frac{4}{3}\right]$

☐ D. $\left(-\infty, -\frac{4}{3}\right] \cup \left[\frac{4}{3}, \infty\right)$

☐ E. $\mathbb{R}$

15.2 Answer Keys

1) A. $4x^5 - 7x + 1$

2) A. $28x + 6$

3) D. $-57 + 7i$

4) C. $\frac{-2x^6}{y}$

5) E. $\left\{x | x \leq \frac{-14}{3}\right\}$

6) A. $x = \frac{\log_3 5}{2}$

7) D. $x = \frac{\log 11 - \log 3}{2 \log 7}$

8) -1.5

9) C. y-intercept

10) B. $2x^3 - x^2 - 3x$

11) B. 4

12) A. -27

13) D.

14) D. $16(m-2)(m+2)$

15) C. A horizontal shift to the right

16) 12

17) A

18) D. $\left(-\frac{3}{2}, -\frac{3}{2}\right)$

19) E.

20) D. 4, 3

21) C. $(x^2 + 2x + 4)(x - 2)$

22) C. 23

23) 42

24) A.

25) D. $n^2 + 2n + 10$

26) D. $0 < y \leq 850$

27) B. $(0, 2)$

28) A. 10

29) B. $\begin{cases} 2x - 2y = 1 \\ x + y = 2 \end{cases}$

30) C. $64 - 16\pi$

31) C. $\{0, 1, 2, 3, 4, 5, 6, 7, 8, 9\}$

32) 20

33) B. $v(x) = 115(1.1)^x$

34) E. $35x^2 + 24xy + 4y^2$

35) C. $g(x) = 8\left(\frac{1}{2}\right)^x$

36) E. $a > b$

37) C. 7.54×10^1

38) A. $\frac{-18b - 7c}{9b}$

39) B.

40) $x = 2$

41) D. 3.91×10^2

42) B. $s = \frac{h-k}{t} + 25t$

43) C. $f(x) = 5(x-3)^2 - 49$

44) A. $y = -16t^2 + 64t + 50$

45) D. $10x + 6xy + 4x^2 + 8xz$

46) D. $\frac{10}{3}$

47) D.

48) A. 4

49) E. 22

50) A. (0,4), (8,0)

51) C. $33cm$

52) D. $x = 2$

53) D. $f^{-1}(x) = \pm\sqrt{x-1}$

54) A. $L(n) = 251.2 + 48.2n$

55) C. $3\sqrt{2}$

56) B. $25,000 < (4,334 - 2,712)t$

57) $(6,0)$, $(3,0)$

58) E.

59) B.

60) E. $\mathbb{R}$

15.3 Answers with Explanation

1) To determine which option is a factor of $12x^8 - 21x^4 + 3x^3$, we first factor out the greatest common factor, which is $3x^3$. This results in $3x^3(4x^5 - 7x + 1)$. Hence $4x^5 - 7x + 1$ will be a factor.

2) To find $f(4x)$, substitute $4x$ into the function: $f(4x) = 3(4x) + 4(4x+1) + 2 = 12x + 16x + 4 + 2 = 28x + 6$, which is option A.

3) To simplify $(-4+9i)(3+5i)$, use the distributive property (FOIL method): $(-4)(3) + (-4)(5i) + (9i)(3) + (9i)(5i) = -12 - 20i + 27i + 45i^2$. Combine these results: $-12 - 20i + 27i - 45 = -57 + 7i$. Therefore, the simplified form is $-57 + 7i$.

4) Calculate $(-2x^2y^2)^3$: $-8x^6y^6$. Multiply by $3x^3y$: $-24x^9y^7$. Divide by $12x^3y^8$: $\frac{-24x^9y^7}{12x^3y^8} = \frac{-2x^6}{y}$. Therefore, the correct answer is option C.

5) To solve the inequality $2x - 4(x+2) \geq x + 6$, first simplify the terms:

$$2x - 4x - 8 \geq x + 6 \Rightarrow -2x - 8 \geq x + 6 \Rightarrow -3x \geq 14 \Rightarrow x \leq \frac{-14}{3}.$$

In set-builder notation, this is $\{x | x \leq \frac{-14}{3}\}$, option E.

6) To solve $3^{2x} = 5$, take the logarithm with base 3 of both sides: $\log_3 3^{2x} = \log_3 5$. This simplifies to $2x \log_3 3 = \log_3 5$. Since $\log_3 3 = 1$, the equation becomes $2x = \log_3 5$. Dividing both sides by 2 gives $x = \frac{\log_3 5}{2}$.

7) Start with $3(7^{2x}) = 11$. Divide both sides by 3 to get $7^{2x} = \frac{11}{3}$. Taking the logarithm of both sides gives $2x \log 7 = \log \frac{11}{3}$, which simplifies to $2x \log 7 = \log 11 - \log 3$. Dividing by $2 \log 7$, we find $x = \frac{\log 11 - \log 3}{2 \log 7}$.

8) Factor the equation: $2x^3 - x^2 - 6x = x(2x^2 - x - 6) = x(x-2)(2x+3) = 0$. The solutions are $x = 0$, $x = 2$, and $x = -1.5$. The negative solution is $x = -1.5$.

9) The y-intercept of a graph is the value of y when $x = 0$. For the equation $y = 3x^2 + 6x + 2$, when $x = 0$, $y = 2$, which is the constant term in the equation. Therefore, the y-intercept is directly displayed as the constant term in the equation.

10) To determine which equation has $2x-3$ as a factor, you can either perform polynomial division or evaluate the equation at the root given by $2x-3=0$, which is $x=\frac{3}{2}$. Substituting $x=\frac{3}{2}$ into each equation, only $2x^3-x^2-3x$ equals zero, indicating that $2x-3$ is a factor of it.

11) To find the rate of change in revenue with respect to the number of gadgets sold, we select two points from the table. Using the first two points, where the number of gadgets sold are 2 hundred and 5 hundred resulting in revenues of 10 and 22 thousand dollars respectively. We calculate the change in revenue and the number of gadgets sold. The change in revenue is $22-10=12$, and the change in the number of gadgets sold is $5-2=3$ (in hundreds). Thus, the rate of change is $\frac{12}{3}=4$ thousand dollars per hundred gadgets, corresponding to option B.

12) The zero of the function $p(x)=\frac{5}{9}x+15$ is the value of x that makes $p(x)=0$. We set the function equal to zero and solve for x:

$$0=\frac{5}{9}x+15\Rightarrow -15=\frac{5}{9}x.$$

Multiplying both sides by $\frac{9}{5}$ to isolate x gives: $x=-27$. Thus, the zero of the function is $x=-27$.

13) In the function $g(x)=8x+150$, x represents the number of books purchased, as $8x$ calculates the total cost of purchasing x books at \$8 dollars each. The term 150 is a fixed amount, representing the annual membership fee. Therefore, x must represent the number of books the member buys in a year, making the total cost a combination of the membership fee and the cost of the books.

14) The expression $16m^2-64$ can be factored using the difference of squares formula, $a^2-b^2=(a-b)(a+b)$. Therefore, the expression can be factored as:

$$16m^2-64=16(m^2-4)=16(m-2)(m+2).$$

15) The transformation from $f(x)=\pi^x$ to $h(x)=\pi^{x-2}$ involves a shift in the x-direction. The subtraction of 2 from x in the exponent indicates a horizontal shift to the right by 2 units.

16) To find the sum of the polynomials $-5x^2+4x-30$ and $9x^2-6x+20$, we combine like terms:

$$(-5x^2+4x-30)+(9x^2-6x+20)=(9x^2-5x^2)+(4x-6x)+(-30+20).$$

Simplifying each group of like terms, we get: $4x^2 - 2x - 10$. Therefore, $a = 4$, $b = -2$, and $c = -10$. The value of $a + b - c$ is:

$$4 - 2 - (-10) = 4 - 2 + 10 = 12.$$

Thus, the value of $a + b - c$ is 12.

17) The function $m(x) = g(x-3) + 2$ is derived from the parent function $g(x) = x^2$ by shifting it 3 units to the right and 2 units up. This transformation corresponds to the function $(x-3)^2 + 2$, which is exactly what is plotted in graph A. Therefore, graph A best represents the function $m(x)$.

18) The vertex of a quadratic function $h(x) = ax^2 + bx + c$ is at the point $\left(-\frac{b}{2a}, h(-\frac{b}{2a})\right)$. For the function $h(x) = 2x^2 + 6x + 3$, $a = 2$ and $b = 6$. Thus, the x-coordinate of the vertex is $-\frac{6}{2\times 2} = -\frac{3}{2}$. Substituting $-\frac{3}{2}$ into the function to find the y-coordinate, we get $h(-\frac{3}{2}) = 2(-\frac{3}{2})^2 + 6(-\frac{3}{2}) + 3 = \frac{9}{2} - 9 + 3 = -\frac{3}{2}$. Thus, the vertex is $\left(-\frac{3}{2}, -\frac{3}{2}\right)$.

19) The value 1.08 in the function represents the growth factor of the investment per year. The base number 1 represents the initial amount (or 100% of the initial amount), and the 0.08 represents an 8% increase from the initial amount each year. Therefore, the account grows at a rate of 8% per year, making option E the correct interpretation.

20) The zeros of the function can be found by factoring the quadratic equation $x^2 - 7x + 12$. We need two numbers whose product is 12 and sum is -7. These numbers are -4 and -3. Therefore, the function can be factored as $(x-4)(x-3) = 0$. Setting each factor equal to zero gives the roots $x = 4$ and $x = 3$, which are the zeros of the function.

21) The binomial $x^3 - 8$ is a difference of cubes, which can be factored using the formula $a^3 - b^3 = (a-b)(a^2 + ab + b^2)$. Here, $a = x$ and $b = 2$. Applying the formula, we get $(x-2)(x^2 + 2x + 4)$ as the factors of $x^3 - 8$.

22) The angle $(5x - 20)^\circ$ must be acute, so $0 < 5x - 20 < 90$. Solving for x, we find: $20 < 5x < 110 \Rightarrow 4 < x < 22$. Thus, $x = 23$ is outside this range, making it not possible for the angle to remain acute.

23) The y-intercept of a function is found by evaluating the function at $x = 0$. For the function $k(x) = 42\left(\frac{4}{5}\right)^x$, substitute $x = 0$ to find the y-intercept: $k(0) = 42\left(\frac{4}{5}\right)^0 = 42 \times 1 = 42$. Therefore, the value of the y-intercept is

42.

24) The constraint for the musician's total gig hours is represented by the inequality $x+y \leq 150$. The graph in option A correctly visualizes this constraint by shading the region below the line $y=150-x$. This line represents all possible values where the sum of hours between the two gigs equals 150, and the shaded area below this line represents all combinations where the total hours are less than 150. The other graphs either do not correctly display the inequality or represent different constraints that do not align with the given problem's conditions.

25) To simplify the expression $(3n^2+2n+6)-(2n^2-4)$, distribute the negative sign to the terms in the second parentheses and then combine like terms:

$$3n^2+2n+6-2n^2+4=(3n^2-2n^2)+2n+(6+4)=n^2+2n+10.$$

So, the expression simplifies to $n^2+2n+10$, which corresponds to option D.

26) The range of a function represents the set of all possible output values (y-values). From the graph, it is evident that the population starts at 850 and decreases exponentially as x increases. The population never reaches zero, hence the inequality $0<y$ (as the population can't be negative or zero). However, the population starts at 850, hence the inequality $y \leq 850$. Combining these gives $0<y \leq 850$, making option D the correct representation of the range.

27) To find which ordered pair satisfies the inequality $3y-4x \geq 6$, substitute the x and y values from each option into the inequality. After checking all options, only $(0,2)$ satisfies the inequality as $3(2)-4(0)=6 \geq 6$. Therefore, the correct answer is B.

28) Given $2^p.4^q=2^{10}$, and knowing that $4^q=(2^2)^q=2^{2q}$, we can combine the exponents on the left side: $2^p \cdot 2^{2q}=2^{p+2q}$. This must equal 2^{10}. Therefore, $p+2q=10$.

29) The correct system of equations is identified by analyzing the graphical and tabular representations of lines h and i. Line i, depicted on a coordinate grid, follows the equation $y=-x+2$, which is a rearrangement of $x+y=2$, matching the second equation of option B.

Line h is represented by a table of values, from which the equation $y=x-0.5$ can be derived by calculating the slope and y-intercept. This equation, when rearranged to $2x - 2y = 1$, aligns with the first equation in option B.

Hence, the correct answer is B. $\begin{cases} 2x - 2y = 1 \\ x + y = 2 \end{cases}$.

30) To find the area of the shaded region, we start by calculating the area of the square and the circle. The square's side length is twice the radius of the circle, giving it a side length of 8. Therefore, the area of the square is: $8^2 = 64$. The circle, with a radius of 4, has an area of: $\pi \times 4^2 = 16\pi$. The shaded region is the area of the square minus the area of the circle: $64 - 16\pi$. So, the correct answer is C. $64 - 16\pi$, representing the area of the square not covered by the circle.

31) The domain of a function is the set of all possible input values (x-values). In this context, the domain represents the number of players on the team, which can range from 0 to 9 players as stated by the coach. Each player is allocated 9 arrows, which is represented by the linear function $y = 9x$ on the graph.

Therefore, the set $\{0, 1, 2, 3, 4, 5, 6, 7, 8, 9\}$ best represents the domain of this function, as it includes all integer numbers of players from 0 to 9, corresponding to the possible team sizes the coach is considering. This matches option C.

32) To find $q(2)$, substitute $x = 2$ into the function $q(x) = 3(x-5)^2 - 7$:

$$q(2) = 3(2-5)^2 - 7 = 3(-3)^2 - 7 = 3(9) - 7 = 27 - 7 = 20.$$

Therefore, the value of $q(2)$ is 20.

33) According to the data in the table, the exponential function model is increasing, therefore, the function must have a base greater than 1. It means that the options A or B are correct. Now, put the values in these functions (A and B) and compare the output values:

Year	Given Value	Option A	Option B
1	129	177.1	126.5
2	141	194.8	139.2
3	150	214.3	153
4	169	235.7	168.4
5	188	259.3	185.2

The information in the table above shows that the option B is more suitable.

34) To expand $(7x+2y)(5x+2y)$, use the distributive property (FOIL method):

$$(7x+2y)(5x+2y) = 7x \cdot 5x + 7x \cdot 2y + 2y \cdot 5x + 2y \cdot 2y = 35x^2 + 14xy + 10xy + 4y^2 = 35x^2 + 24xy + 4y^2.$$

Thus, the expansion results in $35x^2 + 24xy + 4y^2$, which corresponds to option E.

35) Consider the exponential formula $y = ab^x$ and use the given points to determine constants a and b. From the point $(0,8)$, we find $y = 8$ when $x = 0$, indicating the initial value $a = 8$. Using the point $(2,2)$, we solve for b: $2 = 8b^2 \Rightarrow b^2 = \frac{1}{4} \Rightarrow b = \pm\frac{1}{2}$. We discard negative value, because the base cannot be negative. So, we have $b = \frac{1}{2}$. Thus, the function is $g(x) = 8\left(\frac{1}{2}\right)^x$. This corresponds to option C.

36) Substituting $(0,0)$ into the system of inequalities:

For $y < x + a$, we get $0 < 0 + a \Rightarrow a > 0$.

For $y > x + 2b$, we get $0 > 0 + 2b \Rightarrow b < 0$.

Since $a > 0$ and $b < 0$, it is clear that $a > b$, which is option E.

37) To multiply the numbers in scientific notation, multiply the coefficients and add the exponents:

$$(2.9 \times 10^6) \times (2.6 \times 10^{-5}) = (2.9 \times 2.6) \times 10^{6-5} = 7.54 \times 10^1.$$

Therefore, the product is 7.54×10^1, which corresponds to option C.

38) To simplify $-2+\frac{3b-4c}{9b}-\frac{2b+2c}{6b}$, first find a common denominator for the fractions, which is $18b$:

$$\frac{-2(18b)}{18b}+\frac{2(3b-4c)}{18b}-\frac{3(2b+2c)}{18b}=\frac{-36b+6b-8c-6b-6c}{18b}=\frac{-36b-14c}{18b}.$$

Simplifying further, we get $\frac{-18b-7c}{9b}$ after dividing both numerator and denominator by 2. Therefore, the equivalent expression is $\frac{-18b-7c}{9b}$, which corresponds to option A.

39) The quadratic function $y=2x^2-4x-1$ can be visually identified by examining the vertex and direction of the parabola. Since the coefficient of x^2 is positive, the parabola opens upwards. To identify the correct graph, observe:

- The shape (upward opening),
- The vertex position (found by $x=-\frac{b}{2a}$, which in this case is $x=\frac{4}{4}=1$),
- The y-intercept (when $x=0$, $y=-1$).

The graph in option B accurately reflects these characteristics. It opens upwards, the vertex appears at $x=1$, and the y-intercept matches at $y=-1$.

40) To find the solutions, first rearrange the equation: $x^2-3x+1=x-3$ becomes $x^2-4x+4=0$. This equation can be factored as $(x-2)^2=0$. This factorization indicates that there is one repeated solution, $x=2$.

41) To multiply the numbers in scientific notation, multiply the coefficients and add the exponents:

$$(1.7\times 10^9)\times(2.3\times 10^{-7})=(1.7\times 2.3)\times 10^{9-7}=3.91\times 10^2.$$

Therefore, the product is 3.91×10^2, which corresponds to option D.

42) Start with the equation $h=-25t^2+st+k$. To solve for s, rearrange the equation: $st=h+25t^2-k$. Then divide by t to isolate s: $s=\frac{h+25t^2-k}{t}=\frac{h-k}{t}+25t$. So, the correct expression for s in terms of h, t, and k is $s=\frac{h-k}{t}+25t$, which corresponds to option B.

43) To find the equivalent function in vertex form, complete the square for $f(x)=5x^2-30x-4$:

$$f(x)=5(x^2-6x)-4=5[x^2-6x+(3)^2]-5(3)^2-4=5(x-3)^2-45-4=5(x-3)^2-49.$$

So, the equivalent function is $f(x)=5(x-3)^2-49$.

44) To identify which function best represents the graph, analyze the shape and key points of the parabola:
- The graph starts at a height of 50 feet at time $t = 0$ seconds, which must match the constant term in the equation.
- The direction of the parabola is downward, indicating the leading coefficient of t^2 is negative.
- The graph reaches a peak and then decreases, suggesting the presence of a positive linear term in t.

Comparing the given options:
- Option A, $y = -16t^2 + 64t + 50$, starts at 50 feet ($t = 0$), rises due to the positive $64t$ term, and turns downward from the $-16t^2$ term, matching the graph's behavior.
- Other options either do not start at 50 feet, lack the necessary upward motion contributed by a positive linear term, or have incorrect signs that do not match the graph's opening direction.

Thus, option A correctly represents the trajectory.

45) To simplify $2x(5+3y+2x+4z)$, distribute $2x$ to each term inside the parentheses:

$$2x \cdot 5 + 2x \cdot 3y + 2x \cdot 2x + 2x \cdot 4z = 10x + 6xy + 4x^2 + 8xz.$$

This matches option D.

46) To solve the equation, use the logarithm rule $\log_b(a) - \log_b(c) = \log_b\left(\frac{a}{c}\right)$. This gives us $\log_4\left(\frac{x+2}{x-2}\right) = 1$. To remove the logarithm, rewrite the equation in exponential form: $4^1 = \frac{x+2}{x-2}$, which simplifies to $4 = \frac{x+2}{x-2}$. Solve for x:

$$4x - 8 = x + 2 \Rightarrow 3x = 10 \Rightarrow x = \frac{10}{3}.$$

Now substitute $x = \frac{10}{3}$ into the original logarithmic equation:

$$\log_4\left(\frac{10}{3} + 2\right) - \log_4\left(\frac{10}{3} - 2\right) = \log_{2^2}\left(\frac{\frac{10}{3}+2}{\frac{10}{3}-2}\right) = \frac{1}{2}\log_2(4) = \log_2 2 = 1.$$

We find that it satisfies the equation, making D the correct answer.

47) To verify the correct zeros of the polynomial $g(x) = 3x^2 + 20x - 7$, we substitute the zeros suggested by each option into the polynomial and check if they result in zero:

- $g(3) = 3(3)^2 + 20(3) - 7 = 27 + 60 - 7 = 80 \neq 0$
- $g(-7) = 3(-7)^2 + 20(-7) - 7 = 147 - 140 - 7 = 0$

- $g\left(\frac{1}{3}\right) = 3\left(\frac{1}{3}\right)^2 + 20\left(\frac{1}{3}\right) - 7 = \frac{1}{3} + \frac{20}{3} - 7 = 0$
- $g(7) = 3(7)^2 + 20(7) - 7 = 147 + 140 - 7 = 280 \neq 0.$

So, $x = -7$ and $x = \frac{1}{3}$ are zeros and the factors are $(x+7)$ and $(3x-1)$. These match Option D.

48) To find the number of tickets purchased, set the total cost equal to 103.25 and solve for t: $103.25 = 24.50t + 5.25$. Subtract 5.25 from both sides: $98 = 24.50t$. Divide by 24.50 to find t: $t = 4$. So, the customer purchased 4 tickets.

49) Let's denote the number of students as s. With 2 liters per student, Mrs. Johnson has 4 liters left, so the total liquid is $m = 2s + 4$. For 3 liters per student, she needs 18 more liters, implying $m + 18 = 3s$. Substituting m from the first equation into the second gives $2s + 4 + 18 = 3s$, which simplifies to $s = 22$. This matches E.

50) For the x-intercept, set $y = 0$: $2x + 4(0) = 16$, giving $x = 8$. Thus, the x-intercept is $(8,0)$. For the y-intercept, set $x = 0$: $2(0) + 4y = 16$, giving $y = 4$. So, the y-intercept is $(0,4)$. The correct intercepts are $(0,4)$ for y and $(8,0)$ for x.

51) To estimate the perimeter of the shape, consider the perimeters of the semicircles and the radius of the large semicircle. The large semicircle has a diameter of 12 cm, giving a radius of 6 cm. Its perimeter is $P_1 = \frac{2\pi r}{2} \approx \frac{12\times 3}{2} \approx 18$cm. The smaller semicircle has a diameter of 6 cm or radius 3 cm. Its perimeter is $P_2 = \frac{2\pi r}{2} \approx \frac{6\times 3}{2} \approx 9$cm. The total perimeter:

$$P = P_1 + P_2 + \text{ radius of the large semicircle } \approx 18 + 9 + 6 \approx 33cm.$$

52) In the graph of a quadratic function, the axis of symmetry is a vertical line through the vertex of the parabola. The equation of this line is always of the form $x = k$, where k is the x-coordinate of the vertex. Here, the x-coordinate of the vertex is 2, so the correct equation representing the axis of symmetry is $x = 2$.

53) Replace $f(x)$ with y, yielding $y = x^2 + 1$. To find the inverse, we swap x and y and solve for y:

$$x = y^2 + 1 \Rightarrow x - 1 = y^2 \Rightarrow y = \pm\sqrt{x-1}.$$

So, the inverse function is $f^{-1}(x) = \pm\sqrt{x-1}$.

54) The total length includes the lengths of all n sections and the two towers. The towers' combined height adds to the length, so $2 \times 125.6 = 251.2$. The length of the bridge sections is $48.2n$. Therefore, the total length function is $L(n) = 251.2 + 48.2n$.

55) Simplify each term:

$$\sqrt{8} - \sqrt{50} + \sqrt{72} = 2\sqrt{2} - 5\sqrt{2} + 6\sqrt{2} = (2 - 5 + 6)\sqrt{2} = 3\sqrt{2}.$$

56) The savings per year is the difference between the original and new costs: $4,334 - 2,712$. The total savings over t years is this difference times t. The inequality $25,000 < (4,334 - 2,712)t$ represents the condition where the total savings exceed the initial investment of $25,000$.

57) To determine the x-intercepts of the parabola $y = x^2 - 9x + 18$, set $y = 0$ and factorize the quadratic: $0 = (x-3)(x-6)$. Solving for x gives the intercepts: $x = 3$ and $x = 6$. Thus, the x-intercepts are at points $(3,0)$ and $(6,0)$.

58) Observing the coefficients of x and y in both equations, they are multiples of each other ($3x$ and $9x$, $5y$ and $15y$), and the constants on the right-hand side (8 and 24) are also multiples. This indicates that the two equations are actually the same line when graphed, leading to an infinite number of solutions where the two equations intersect, which is every point on the line itself.

59) The axis of symmetry for a quadratic function $ax^2 + bx + c$ is given by $x = -\frac{b}{2a}$:
- For $f(x)$, with $a = 2$ and $b = 0$, the axis of symmetry is $x = 0$.
- For $g(x)$, with $a = -1$ and $b = 0$, the axis of symmetry is also $x = 0$.
- For $h(x)$, with $a = 1$ and $b = 1$, the axis of symmetry is $x = -\frac{1}{2}$.

Here, both $f(x)$ and $g(x)$ share the same axis of symmetry, $x = 0$. So, the correct option is B.

60) The domain of any quadratic function is always all real numbers, $\mathbb{R}$, because you can input any real number x into the function and get a real number output. Thus, the domain of $f(x) = -3x^2 + 16$ is $\mathbb{R}$.

16. Practice Test 2

16.1 Practices

1) At a rate of $5d+7$ Kilometers per hour, how many kilometers can a car travel in 6 hours?

☐ A. $10d+14$

☐ B. $15d+21$

☐ C. $30d+42$

☐ D. $20d+28$

☐ E. $35d+49$

2) The regular price for a book is \$50. A bookstore offers a 15% discount on the regular price. What would be the savings in dollars if you purchase 5 books from the discounted bookstore instead of the regular price?

☐ A. \$7.50

☐ B. \$37.50

☐ C. \$75

☐ D. \$150

☐ E. \$375

3) Which of the following is equal to x^{ab} for all values of x, a and b?

☐ A. $x^{(a+b)}$

☐ B. $x^a b$

☐ C. $x^a x^b$

☐ D. $(x^a)^b$

☐ E. xa^b

4) Which expression is equivalent to $\frac{2.8\times10^{-6}}{4\times10^{-9}}$?

☐ A. 7×10^{-1}

☐ B. 7×10^{0}

☐ C. 7×10^{2}

☐ D. 7×10^{3}

☐ E. 7×10^{4}

5) If z varies directly with w, the relationship can be represented by the equation $z = kw$, where k is the proportionality constant. Given that $z = 8$ when $w = 16$, what is the equation of the direct variation that represents this relationship?

☐ A. $z = 0.5w$

☐ B. $z = 2w$

☐ C. $z = 8w$

☐ D. $z = 16w$

☐ E. $z = 128w$

6) If $x = 4\left(\log_3 \frac{1}{81}\right)$, what is the value of x?

☐ A. -16

☐ B. $-\frac{4}{9}$

☐ C. $\frac{4}{4}$

☐ D. $\frac{9}{4}$

☐ E. 16

7) Which table shows the same rate of change of z with respect to w as $z = \frac{4}{3}w - 4$?

☐ A.

w	0	3	6	9
z	-3	1	4	8

☐ B.

w	1	2	3	4
z	$-\frac{8}{3}$	$-\frac{4}{3}$	0	$\frac{4}{3}$

☐ C.

w	-1	0	1	2
z	$-\frac{8}{3}$	-2	$-\frac{8}{3}$	$-\frac{4}{3}$

☐ D.

w	3	6	9	12
z	1	$\frac{16}{3}$	8	$\frac{27}{3}$

☐ E.

w	4	8	12	16
z	$-\frac{4}{3}$	$\frac{8}{3}$	$\frac{20}{3}$	12

8) Which set of ordered pairs contains only points that are on the graph of the function $y = -3x + 6$?

☐ A. $\{(2,-2),(1,4),(3,-3)\}$

☐ B. $\{(4,-8),(2,0),(0,6)\}$

☐ C. $\{(1,3),(2,-1),(-1,9)\}$

☐ D. $\{(-2,12),(0,6),(1,3)\}$

☐ E. $\{(3,3),(5,-9),(-1,9)\}$

9) A graph of a quadratic function is shown on the grid. Which coordinates best represent the vertex of the graph?

☐ A. $(0,2)$

☐ B. $(0,-1)$

☐ C. $(-3,2)$

☐ D. $(-3,0)$

☐ E. $(-5,0)$

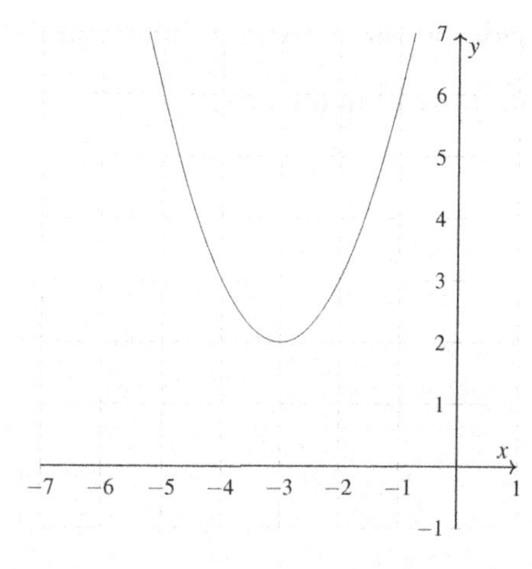

10) The table represents different values of function $f(x)$. What is the value of the expression $2f(-3) - 3f(4)$?

☐ A. -11

☐ B. -9

☐ C. 0

☐ D. 9

☐ E. 11

x	$f(x)$
-4	2
-3	-1
0	0
4	3
5	-4

11) If a quadratic function with equation $y = bx^2 + 7x + 14$, where b is a constant, passes through the point $(3,\ 20)$, what is the value of b^2?

☐ A. 3

☐ B. $\frac{22}{9}$

☐ C. $\frac{25}{9}$

☐ D. 4

☐ E. $\frac{29}{3}$

12) The function $y = 5x + 10$ represents the cost of renting a tool for x hours. Based on this information, which statement about the graph of this situation is true?

☐ A. The y-intercept of the graph represents the cost of each hour of rental.

☐ B. The y-intercept of the graph represents the initial rental cost.

☐ C. The slope of the graph represents the total number of tools rented.

☐ D. The slope of the graph represents the initial rental cost.

☐ E. The graph has one extremum point.

13) The graph of quadratic function f is shown on the grid. If $h(x) = x^2$ and $f(x) = h(x) + m$, what is the value of m? Write your answer in the box. ☐

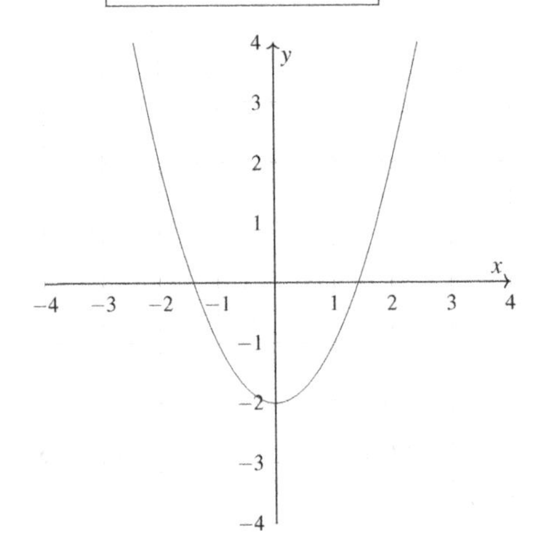

14) In the following equation, what is the value of x?

$$5x - 2 = 3x + 8.$$

☐ A. 2

☐ B. 5

☐ C. 8

☐ D. 10

☐ E. 20

15) Which function is the inverse of $f(x) = \ln \sqrt{x^3}$?

☐ A. e^x

☐ B. x^3

☐ C. $e^{\frac{2x}{3}}$

☐ D. e^{3x}

☐ E. e^{x^3}

16) If $(x-3)^3 = 64$ which of the following could be the value of $(x-7)(x-5)$?

☐ A. 0

☐ B. 12

☐ C. 14

☐ D. 16

☐ E. 18

17) Which graph best represents $y = 3\left(\frac{1}{2}\right)^x$?

☐ A.

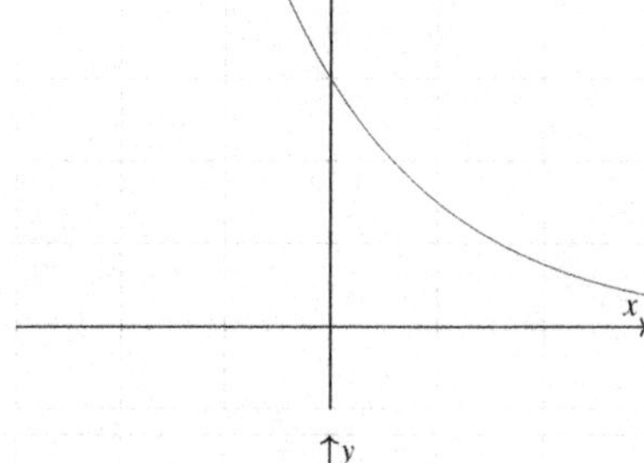

☐ B.

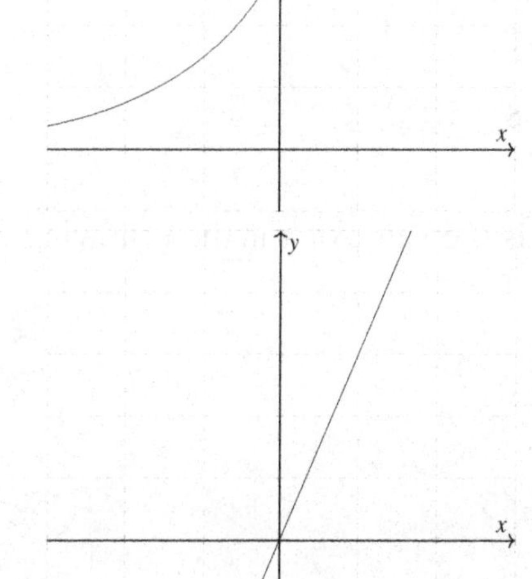

☐ C.

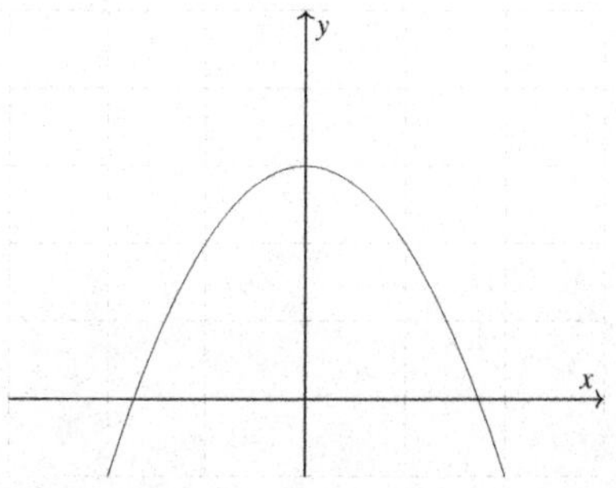

☐ D.

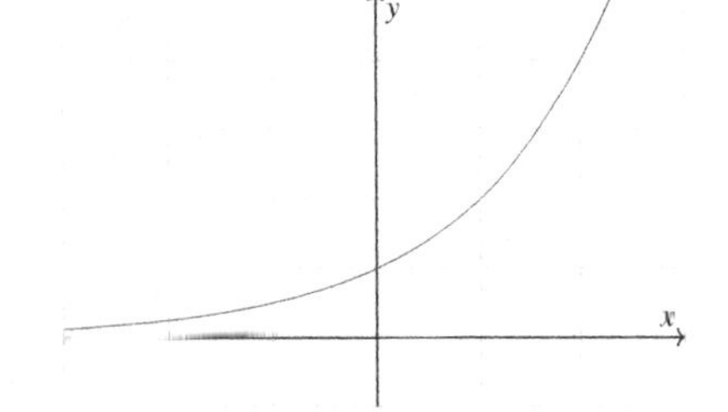

18) The area of an equilateral triangle with sides of length s is:

☐ A. $\frac{s^2}{4}$

☐ B. $\frac{\sqrt{2}\,s^2}{4}$

☐ C. $\frac{\sqrt{3}s^2}{4}$

☐ D. $\frac{s^2}{2}$

☐ E. $\frac{\sqrt{3}s^2}{2}$

19) What is the solution of the following inequality? $|x+5| \leq 7$

☐ A. $-12 \leq x \leq 12$

☐ B. $-2 \leq x \leq 12$

☐ C. $-12 \leq x \leq 2$

☐ D. $-2 \leq x \leq 12$

☐ E. $-2 \leq x \leq 2$

20) Express in scientific notation: 0.0048.

☐ A. 4.8×10^3

☐ B. 4.8×10^2

☐ C. 4.8×10^{-3}

☐ D. 4.8×10^{-4}

☐ E. 4.8×10^{-5}

21) What is the value of x in the following equation?

$$\frac{5}{6}(x-3) = 5\left(\frac{1}{5}x-2\right).$$

☐ A. -45

☐ B. $\frac{3}{5}$

☐ C. $-\frac{3}{5}$

☐ D. -15

☐ E. 45

22) Which of the following is a factor of $9x^2 - 3x - 12$?

☐ A. $x-4$

☐ B. $3x+4$

☐ C. $3x-4$

☐ D. $9x+12$

☐ E. $x-\frac{4}{3}$

23) Simplify the expression $\left(xy^{-2}\right)^3 \left(\frac{y}{x}\right)^{-9}$ to the form $x^n y^m$. What is the value of $n-m$? Write your answer in the box. ☐

24) In the equation below, if c is negative and e is positive, which of the following must be true?

$$\frac{c-e}{ac} = 1.$$

☐ A. $a < 1$

☐ B. $a < -1$

☐ C. $a = 0$

☐ D. $a > -1$

☐ E. $a > 1$

25) Which inequality is equivalent to $6x - 4y < 3y + 42$?

☐ A. $y > \frac{6}{7}x - 7$

☐ B. $y > \frac{6}{7}x - 6$

☐ C. $y > 7 - \frac{6}{7}x$

☐ D. $y < 7 - \frac{6}{7}x$

☐ E. $y < 2x - 7$

26) For what value of x is the function $g(x)$ below undefined?

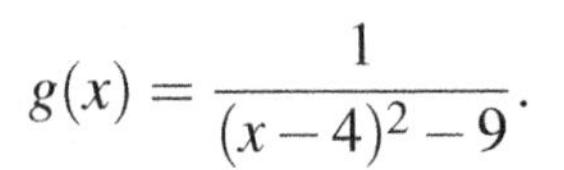

$$g(x) = \frac{1}{(x-4)^2-9}.$$

☐ A. 7

☐ B. −7

☐ C. 1

☐ D. $x = 7$ and $x = 1$

☐ E. $x = -7$ and $x = 1$

27) What is the sum of all values of m that satisfies $3m^2 + 18m + 27 = 0$?

☐ A. 9

☐ B. 6

☐ C. 0

☐ D. −6

☐ E. −9

28) A gardener can plant a tree in 4 hours and a bush in 2 hours. The function below can be used to find the number of trees the gardener plants when he completes g bushes in a 40-hour workweek.

$$t = \frac{(80-2g)}{0.5}.$$

If the gardener planted 8 trees in one week, how many bushes did he complete that week?

☐ A. 18

☐ B. 38

☐ C. 14

☐ D. 28

☐ E. 48

29) Function f is in the form $y = bx^2 + d$. If the value of b is positive and d is negative, which graph could represent f?

☐ A.

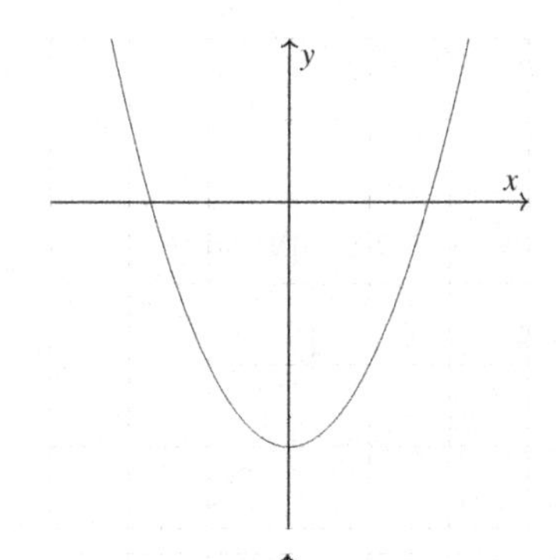

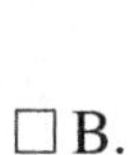

☐ B.

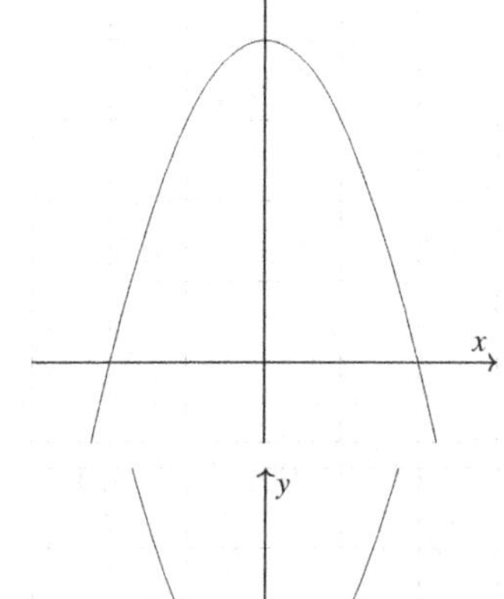

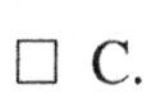

☐ C.

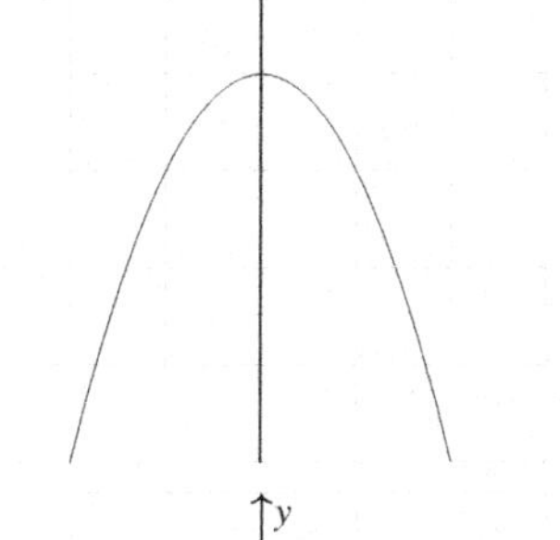

☐ D.

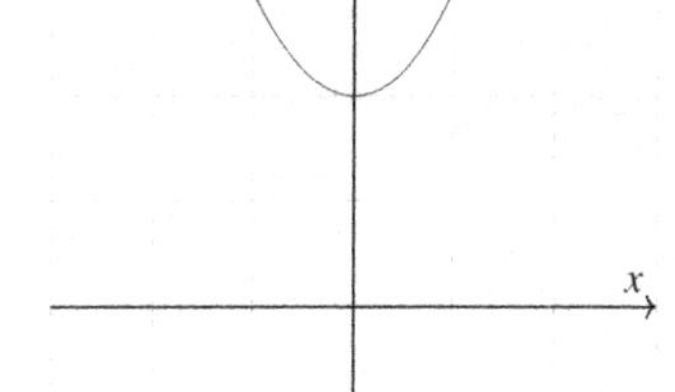

☐ E.

30) The equation below gives the height h, in centimeters, of an object t seconds after it is thrown straight up with an initial speed of v centimeters per second from a height of s centimeters. Which of the following gives v in terms of h, t, and s?

$$h = -t^2 + vt + 2s.$$

☐ A. $v = \frac{h-2s}{t} + t$

☐ B. $v = h - t^2 - 2s$

☐ C. $v = h + t^2 - 2s$

☐ D. $v = h + s - t$

☐ E. $v = \frac{h+t^2+2s}{t}$

31) The graph of $f(x) = x^2$ was transformed to create the graph of $g(x) = f(x) + 3$.

Which of the following statements is true about the graphs of f and g?

☐ A. The graph of g is a reflection of the graph of f across the x-axis.

☐ B. The vertex of the graph of g is 3 units below the vertex of the graph of f.

☐ C. The graph of g is a reflection of the graph of f across the y-axis.

☐ D. The y-intercept of the graph of g is 3 units above the y-intercept of the graph of f.

☐ E. The graph of g is stretched vertically by a factor of 3 compared to the graph of f.

32) Subtract $7x^2 - 4$ from triple the quantity $-2x^2 - 3x + 6$.

☐ A. $-6x^2 + 9x + 22$

☐ B. $-13x^2 - 9x + 22$

☐ C. $13x^2 + 9x + 2$

☐ D. $6x^2 - 9x - 22$

☐ E. $-5x^2 - 9x - 18$

33) If $(x-3)^2 + 2 > 4x + 6$, then x can equal which of the following?

☐ A. 0

☐ B. 1

☐ C. 2

☐ D. 3

☐ E. 4

34) Simplify $(4x-6)^2$.

☐ A. $16x^2 + 48x + 36$

☐ B. $16x^2 - 48x + 36$

☐ C. $16x^2 - 48x - 36$

☐ D. $16x^2 + 36$

☐ E. $16x^2 - 36$

35) The table shows the number of items sold by a company every month for nine months. An exponential function can be used to model the data. Which function best models the data?

Time, x (months)	Items Sold, s(x)
1	150
2	165
3	182
4	200
5	220
6	242
7	267
8	294
9	324

☐ A. $s(x) = 0.90(149.5)^{x-1}$

☐ B. $s(x) = 1.10(150)^{x-1}$

☐ C. $s(x) = 150(1.10)^{x-1}$

☐ D. $s(x) = 150(0.90)^{x-1}$

☐ E. $s(x) = 149.5(1.10)^{x-1}$

36) Given $h(x) = x^2 - 6x - 7$, which statement is true?

☐ A. The zeroes are 7 and 1, because the factors of h are $(x-7)$ and $(x-1)$.

☐ B. The zeroes are 7 and -1, because the factors of h are $(x-7)$ and $(x+1)$.

☐ C. The zeroes are -7 and 1, because the factors of h are $(x+7)$ and $(x-1)$.

☐ D. The zeroes are -7 and -1, because the factors of h are $(x+7)$ and $(x+1)$.

☐ E. The zeroes are 7 and 1, because the factors of h are $(x-7)$ and $(x+1)$.

37) The graph of a function is shown on the grid. Which ordered pair best represents the location of the y-intercept?

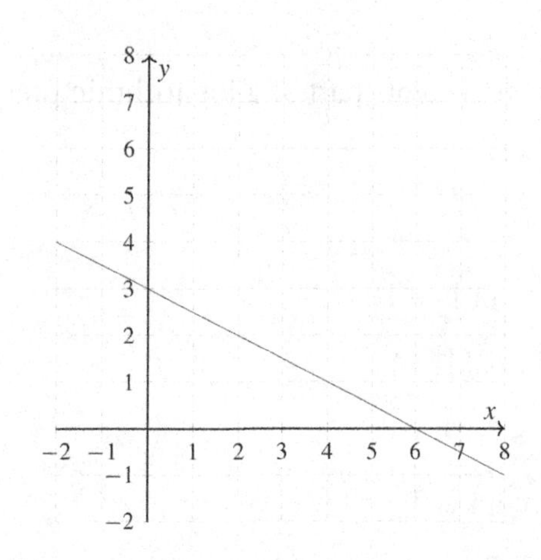

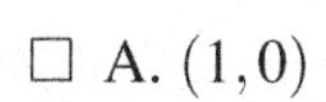

☐ A. $(1,0)$

☐ B. $(0,3)$

☐ C. $(2,0)$

☐ D. $(0,-3)$

☐ E. $(-3,0)$

38) The graph of an exponential function is shown on the grid. Which dashed line is an asymptote for the graph.

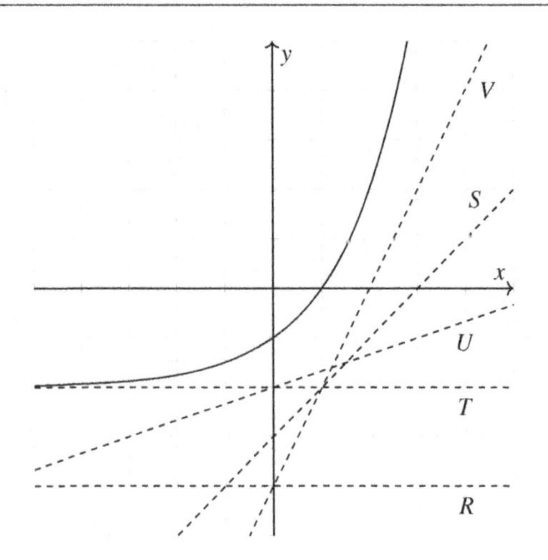

☐ A. Line R

☐ B. Line S

☐ C. Line T

☐ D. Line U

☐ E. Line V

39) Which function is equivalent to $g(x) = 3(3-4x)^2 - 9$?

☐ A. $g(x) = 36x^2 + 48x$

☐ B. $g(x) = 48x^2 - 9$

☐ C. $g(x) = 36x^2 + 48x - 9$

☐ D. $g(x) = 48x^2 - 72x + 18$

☐ E. $g(x) = -48x^2 - 48x - 9$

40) The graph below represents part of a logarithmic function. Which statement is best supported by the graph?

☐ A. The y-intercept is -1.

☐ B. The domain is $(1, \infty)$.

☐ C. The range is $(-\infty, 0)$.

☐ D. The x-intercept is 1.

☐ E. The range is $\mathbb{R}^+$.

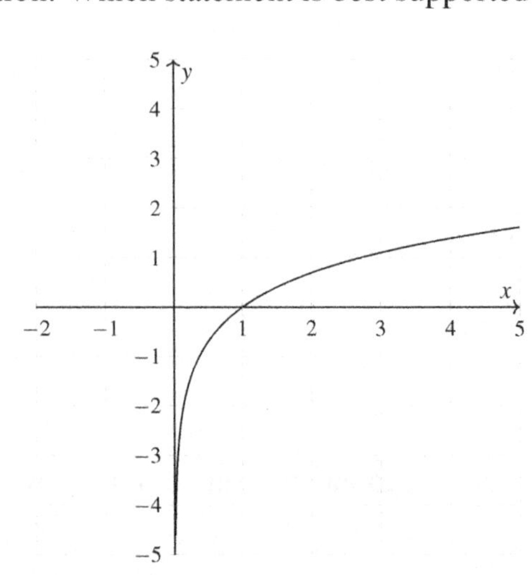

41) The graph of the quadratic function h was transformed to create the graph of $j(x) = h(x-2) - 1$.

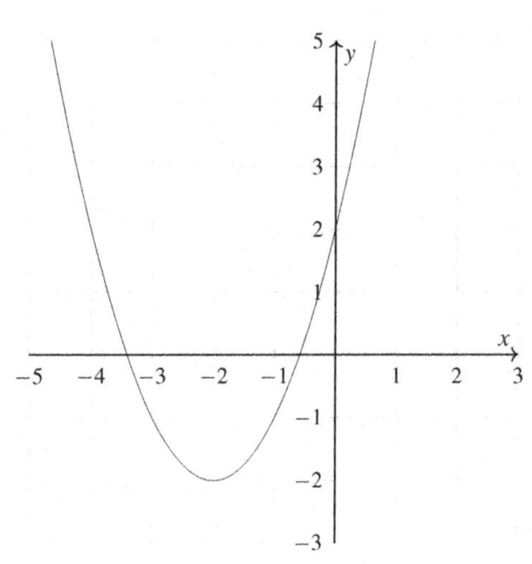

Which graph best represents j?

☐ A.

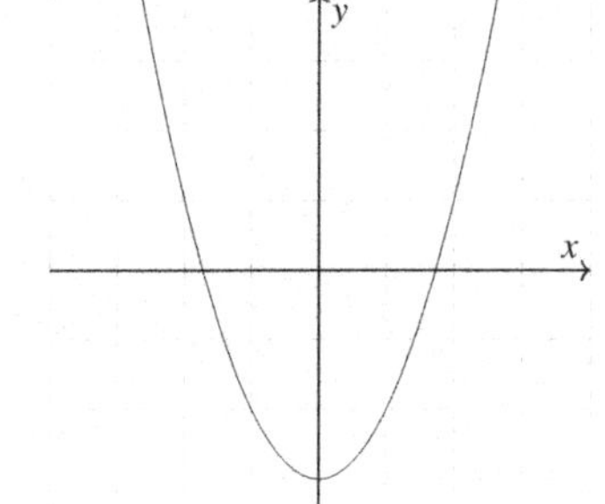

☐ B.

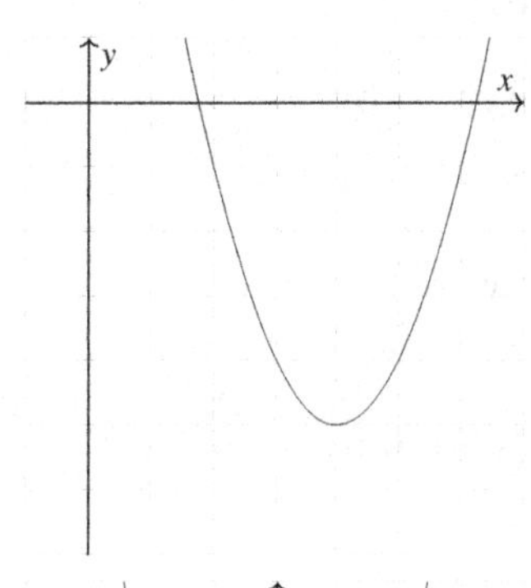

☐ C.

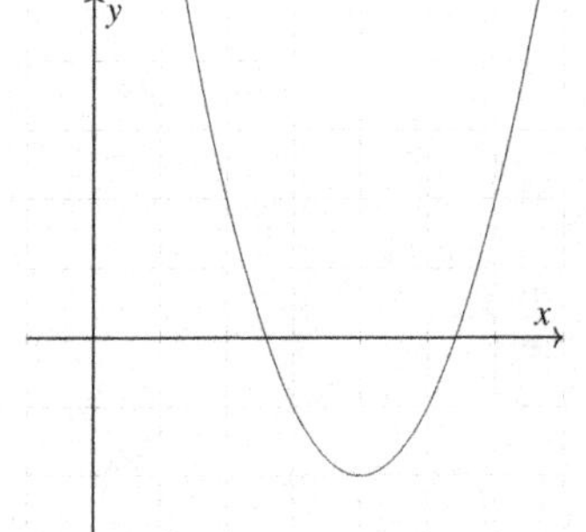

☐ D.

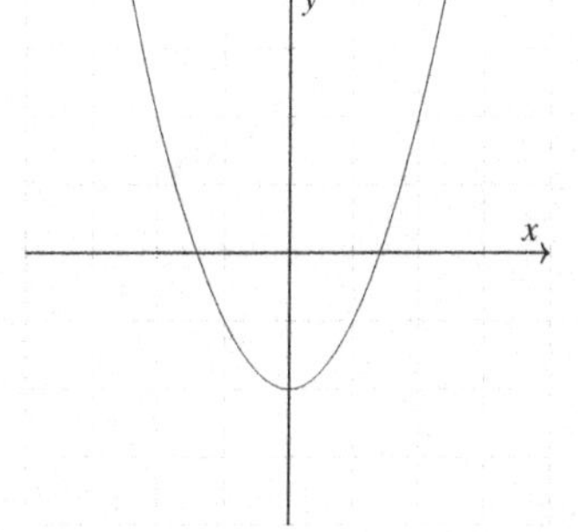

☐ E.

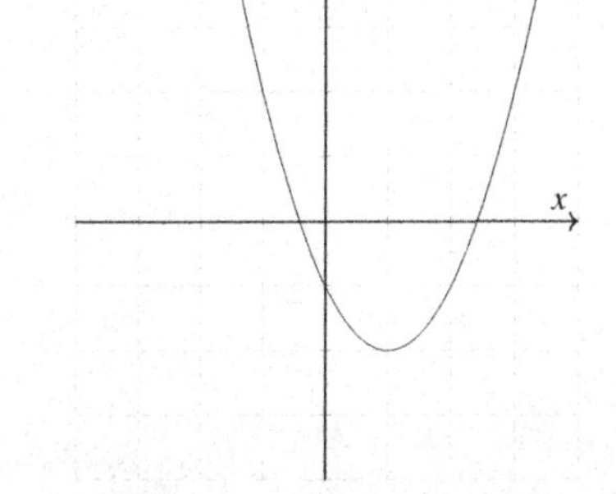

42) In the following figure, point O is the center of the circle and the equilateral triangle has a perimeter of 36.

What is the circumference of the circle? ($\pi = 3.14$) Write your answer in the box. ☐

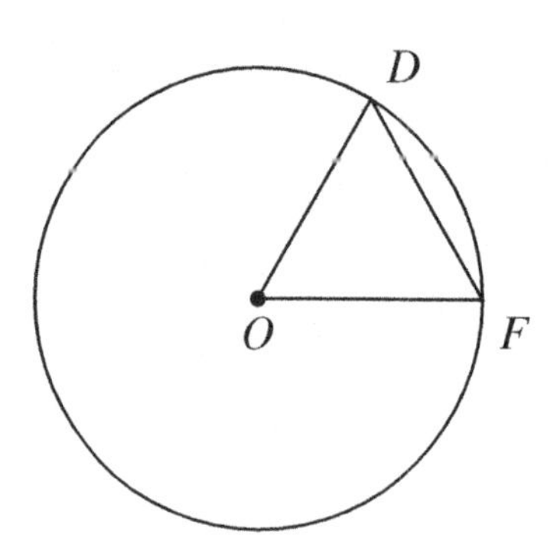

43) Simplify the expression $(3x-4y)^2$.

☐ A. $9x^2 - 16y^2$

☐ B. $9x^2 + 16y^2$

☐ C. $9x^2 - 24xy + 16y^2$

☐ D. $9x^2 - 12xy + 16y^2$

☐ E. $9x^2 - 16xy + 16y^2$

44) Solve this equation for x: $e^{2x} = 25$

☐ A. 25

☐ B. $2\ln(25)$

☐ C. $\ln(25)$

☐ D. $\frac{\ln(25)}{2}$

☐ E. $\frac{e^{2x}}{4}$

45) Simplify $\frac{5-2i}{3i}$.

☐ A. $\frac{2i}{3}$

☐ B. i

☐ C. $i - \frac{2}{3}$

☐ D. $-\frac{2}{3} - \frac{5}{3}i$

☐ E. $\frac{2}{3} - \frac{5}{3}i$

46) The graph of a cubic function is shown on the grid. Which function is best represented by this graph?

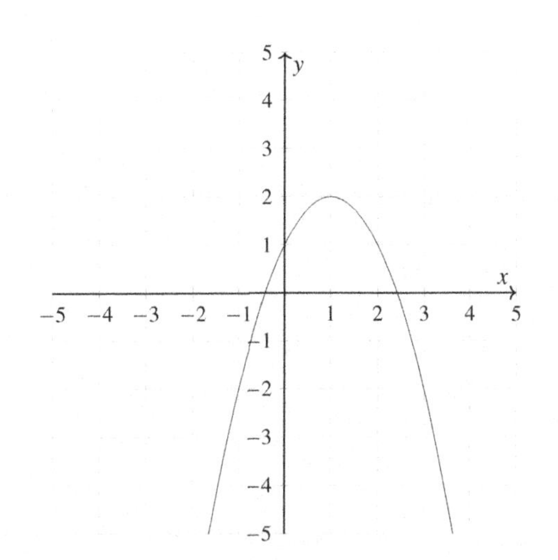

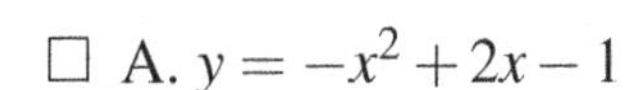

☐ A. $y = -x^2 + 2x - 1$

☐ B. $y = -x^2 - 2x - 1$

☐ C. $y = -x^2 - 2x + 1$

☐ D. $y = -x^2 + 2x + 1$

☐ E. $y = x^2 - 2x + 1$

47) What is the solution to $2(n-1) = 3(n+2) - 10$? Write your answer in the box. ☐

48) If a and b are solutions of the following equation, which of the following is the ratio $\frac{a}{b}$? $(a > b)$

$$2x^2 - 11x + 8 = -3x + 18.$$

☐ A. 5

☐ B. $\frac{1}{5}$

☐ C. $-\frac{1}{5}$

☐ D. -1

☐ E. -5

49) Which statement about the graph of $f(x) = 4(1.25)^x$ is true?

☐ A. The graph includes the point $(-1, 2)$.

☐ B. There is an asymptote to the equation of $x = 0$.

☐ C. The x-intercept is 4.

☐ D. The axis of symmetry is the line with the equation $y = 0$.

☐ E. It is an increasing function.

50) If $x^2 - 16 = 9$, which of the following could be the value of $(x-6)(x-3)$?

☐ A. 2

☐ B. 1

☐ C. 0

☐ D. -1

☐ E. -2

51) Which of the following graphs best shows a strong positive association between x and y?

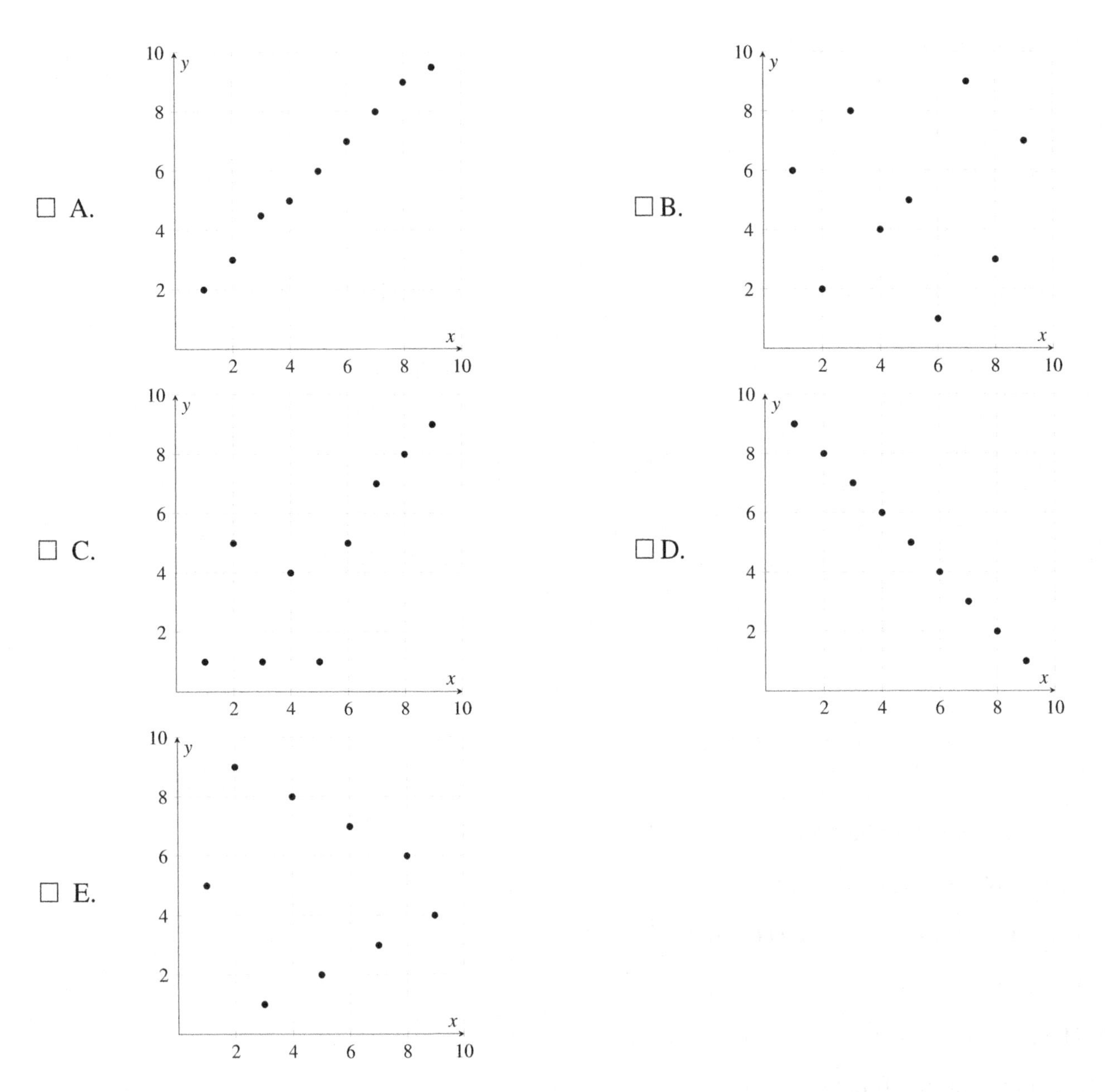

52) Point B lies on the line with equation $y+2=3(x-4)$. If the x-coordinate of B is 5, what is the y-coordinate of B?

☐ A. 1

☐ B. 9

☐ C. 12

☐ D. 15

☐ E. 18

53) The graph of $y = -2x^2 + 8x + 4$ is shown below. If the graph crosses the y-axis at the point $(0, s)$, what is the value of s?

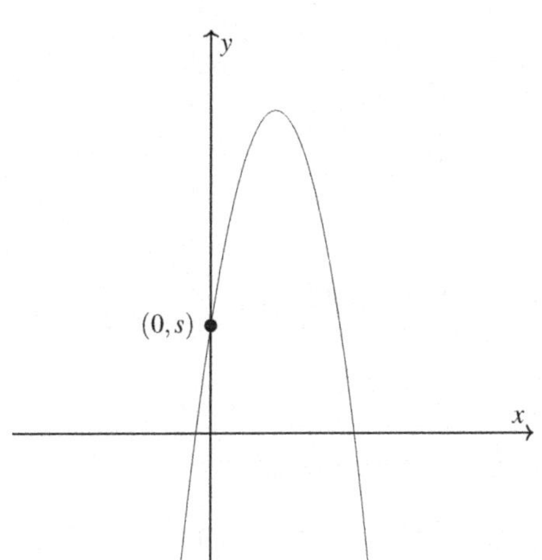

☐ A. 1

☐ B. 2

☐ C. 3

☐ D. 4

☐ E. 5

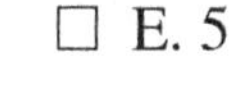

54) What is the range of the function graphed on the grid?

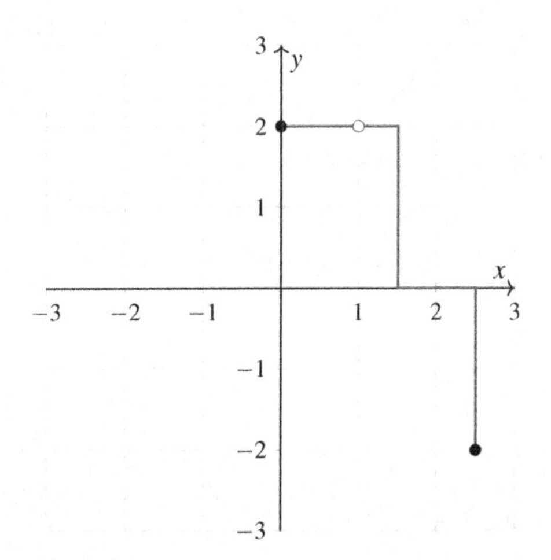

☐ A. $\{y| -2 \leq y \leq 2\}$

☐ B. $\{y| -2 < y \leq 2\}$

☐ C. $\{y| -2 \leq y < 2\}$

☐ D. $\{y| -2 < y < 2\}$

☐ E. $\{y| 0 \leq y \leq 2\}$

55) The first year Eleanor organized a fund-raising event, she invited 50 people. For each of the next 5 years, she invited double the number of people she had invited the previous year. If $f(n)$ is the number of people invited to the fund-raiser n years after Eleanor began organizing the event, which of the following statements best describes the function f?

☐ A. The function f is a decreasing linear function.

☐ B. The function f is an increasing linear function.

☐ C. The function f is a constant function.

☐ D. The function f is a decreasing exponential function.

☐ E. The function f is an increasing exponential function.

56) If the system of inequalities $y < 2x + 1$ and $y \geq -x$ is graphed in the xy-plane below, which quadrant contains no solutions to the system?

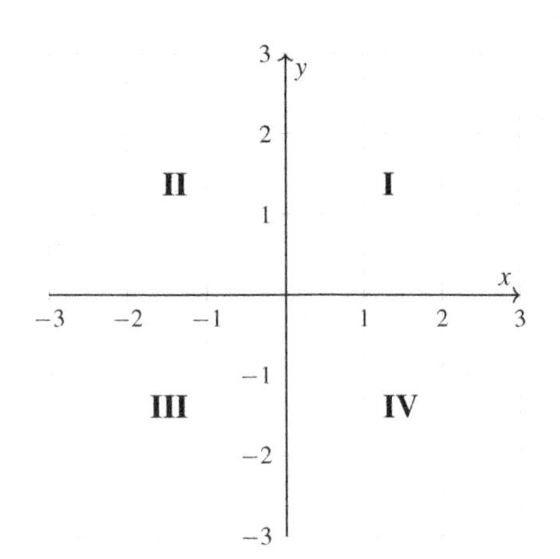

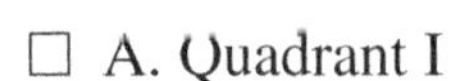
☐ A. Quadrant I

☐ B. Quadrant II

☐ C. Quadrant III

☐ D. Quadrant IV

☐ E. There are solutions in all quadrants.

57) Which situation can be represented by $y = 5x + 7$?

☐ A. The number of cookies, y, in x dozens of cookies after 7 cookies are added to each dozen.

☐ B. The number of people, y, in x groups of 5 people each after adding 7 more people.

☐ C. The cost, y, after a \$7 discount, of buying x jackets that sell for \$5 each.

☐ D. The number of inches, y, in an x-foot-tall tree after adding 7 inches to its height.

☐ E. Choices B and C are correct.

58) Which expression is equivalent to $64x^2 - 16$?

☐ A. $(8x + 4)(8x - 4)$

☐ B. $(8x + 4)(8x + 4)$

☐ C. $(-8x - 4)(8x - 4)$

☐ D. $(8x + 4)(-8x + 4)$

☐ E. $(8x + 4)(-4x + 8)$

59) If $g(x) = 4x + 3(x + 2) + 5$, then $g(2x) = ?$

☐ A. $14x + 11$

☐ B. $17x - 6$

☐ C. $22x + 5$

☐ D. $11x + 10$

☐ E. $14x - 4$

60) Linear function f has an x-intercept of 1 and a y-intercept of -4. Which graph best represents f?

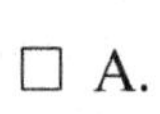
□ A. 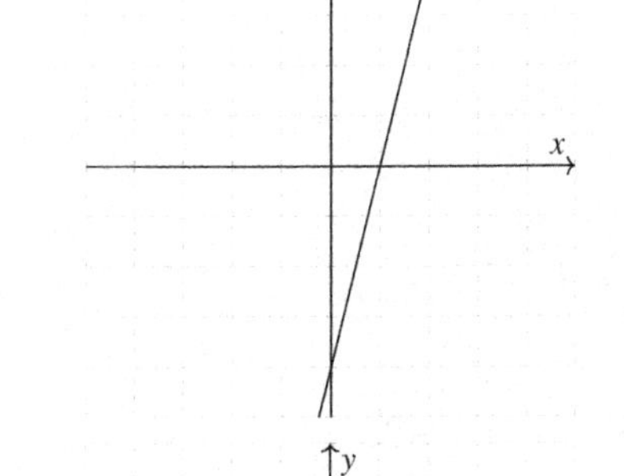

□ B. 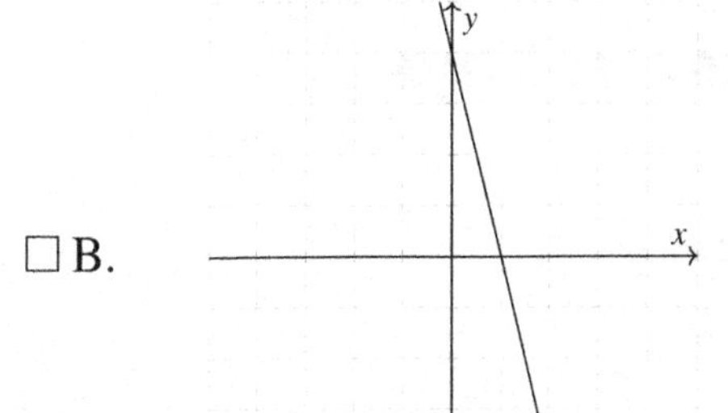

□ C. 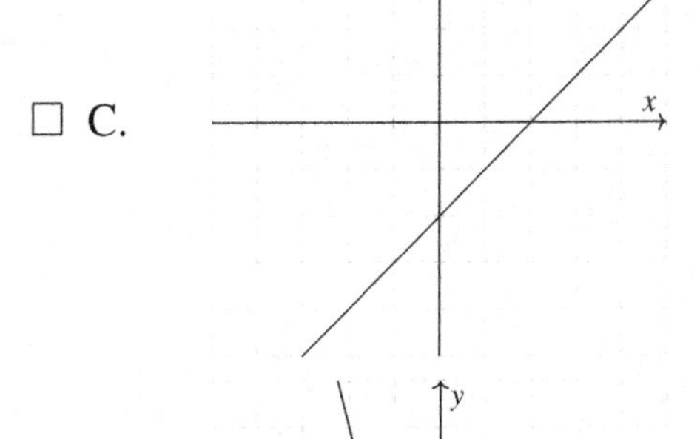

□ D. 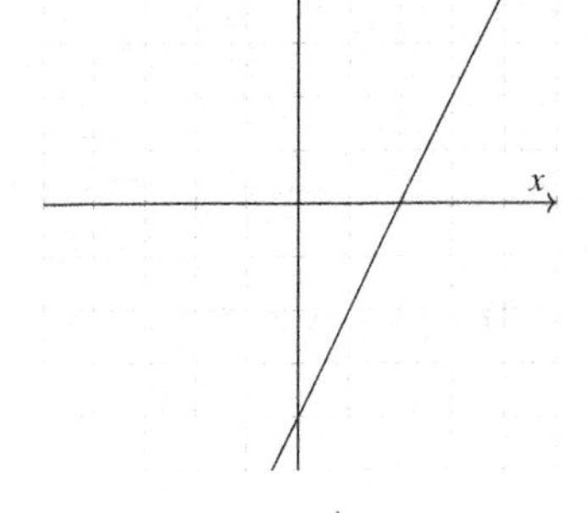

□ E.

16.2 Answer Keys

1) C. $30d+42$

2) B. $37.50

3) D. $(x^a)^b$

4) C. 7×10^2

5) A. $z=0.5w$

6) A. -16

7) B.

8) D. $\{(-2,12),(0,6),(1,3)\}$

9) C. $(-3,2)$

10) A. -11

11) C. $\frac{25}{9}$

12) B.

13) $m=-2$

14) B. 5

15) C. $e^{\frac{2x}{3}}$

16) A. 0

17) A.

18) C. $\frac{\sqrt{3}s^2}{4}$

19) C. $-12\le x\le 2$

20) C. 4.8×10^{-3}

21) E. 45

22) C. $3x-4$

23) $n-m=27$

24) E. $a>1$

25) B. $y>\frac{6}{7}x-6$

26) D. $x=7$ and $x=1$

27) C. -6

28) B. 38

29) A.

30) A. $v=\frac{h-2s}{t}+t$

31) D.

32) B. $-13x^2-9x+22$

33) A. 0

34) B. $16x^2-48x+36$

35) C. $s(x)=150(1.10)^{x-1}$

36) B.

37) B. $(0,3)$

38) C. Line T

39) D. $g(x)=48x^2-72x+18$

40) D. The x-intercept is 1.

41) A.

42) 75.36

43) C. $9x^2-24xy+16y^2$

44) D. $\frac{\ln(25)}{2}$

45) D. $-\frac{2}{3}-\frac{5}{3}i$

46) D. $y=-x^2+2x+1$

47) $n=2$

48) E. -5

49) E.

50) E. -2

51) A.

52) A. 1

53) D. 4

54) A. $\{y|-2\le y\le 2\}$

55) E.

56) C. Quadrant III

57) B.

58) A. $(8x+4)(8x-4)$

59) A. $14x+11$

60) A.

16.3 Answers with Explanation

1) To find the distance a car can travel in 6 hours at a rate of $5d+7$ km/h, we multiply the speed by the time. This is because distance is the product of speed and time. So, we calculate $(5d+7)\times 6 = 30d+42$. This means the car can travel $30d+42$ kilometers in 6 hours, making option C the correct answer.

2) To calculate the savings for purchasing 5 books with a 15% discount, first determine the discount per book: $50\times 0.15 = \$7.50$. This is the savings on one book. Since we are buying 5 books, we multiply the savings per book by 5: $7.50\times 5 = \$37.50$. This total represents the savings when purchasing 5 books at a 15% discount, hence option B is correct.

3) The expression x^{ab} can be interpreted in terms of exponentiation rules. According to the rules of exponents, when you raise a power to another power, you multiply the exponents. So, $(x^a)^b = x^{a\times b} = x^{ab}$. This matches option D.

4) To simplify the expression $\frac{2.8\times 10^{-6}}{4\times 10^{-9}}$, divide the coefficients: $\frac{2.8}{4} = 0.7$. For the exponential part, subtract the exponents of 10: $-6-(-9) = 3$. Thus, the expression simplifies to 0.7×10^3. Since we typically want the coefficient to be a whole number in scientific notation, we adjust 0.7×10^3 to $7\times 10^{-1}\times 10^3 = 7\times 10^2$. Therefore, the correct answer is C.

5) To find the proportionality constant k in the direct variation equation $z = kw$, we use the given values: $z = 8$ when $w = 16$. Plugging these values into the equation gives $8 = k\times 16$. Solving for k, we divide both sides by 16, yielding $k = \frac{8}{16} = 0.5$. Therefore, the equation representing the direct variation is $z = 0.5w$, making option A the correct answer.

6) First, evaluate $\log_3 \frac{1}{81}$. Since $81 = 3^4$, $\log_3 \frac{1}{81} = \log_3 3^{-4} = -4\log_3 3 = -4\times 1 = -4$. Then, multiply by 4: $x = 4\times(-4) = -16$. Thus, the value of x is -16, making option A the correct answer.

7) To determine which table represents the same rate of change as $z = \frac{4}{3}w - 4$, calculate the change in z for a unit change in w. In the equation, the coefficient of w ($\frac{4}{3}$) is the rate of change. Option B shows this same rate: for each increase of 1 in w, z increases by $\frac{4}{3}$. For example, when w increases from 2 to 3, z increases from $-\frac{4}{3}$ to 0, which is an increase of $\frac{4}{3}$. Thus, option B correctly represents the rate of change of z with respect to w.

8) To find which set of ordered pairs lies on the line $y = -3x + 6$, plug in the x-values from each pair into the equation and check if the resulting y-value matches the one in the pair. For set D, plugging $x = -2$ gives $y = -3(-2) + 6 = 12$, $x = 0$ gives $y = -3(0) + 6 = 6$, and $x = 1$ gives $y = -3(1) + 6 = 3$. All these results match the y-values in the set, confirming that all points in set D are on the graph.

9) The vertex form of a quadratic function is the highest/lowest point on the graph. We can see that the vertex of this quadratic function is clearly an ordered pair with coordinates $(-3, 2)$ which matches with option C.

10) Using the table, find the values for $f(-3)$ and $f(4)$, which are -1 and 3 respectively. Then compute the expression: $2f(-3) - 3f(4) = 2(-1) - 3(3) = -2 - 9 = -11$. Therefore, the value of the expression is -11, which corresponds to option A.

11) Substitute the point $(3, 20)$ into the equation $y = bx^2 + 7x + 14$ to find b. This gives $20 = b(3)^2 + 7(3) + 14$. Solving for b, we get $20 = 9b + 21 + 14$. Simplifying, $20 = 9b + 35$, then $9b = -15$, so $b = -\frac{5}{3}$. Squaring b gives $b^2 = \left(-\frac{5}{3}\right)^2 = \frac{25}{9}$, hence option C is the correct answer.

12) In the linear function $y = 5x + 10$, 5 represents the cost per hour (slope), and 10 represents the initial cost (y-intercept), making option B the correct answer.

13) Notice that for the function $h(x)$, if $m > 0$ then $f(x) = h(x) + m$ is shifted up by $|m|$ units, and if $m < 0$ the function $f(x) = h(x) + m$ is shifted down by $|m|$ units. The graph of the function $h(x) = x^2$ is as follows:

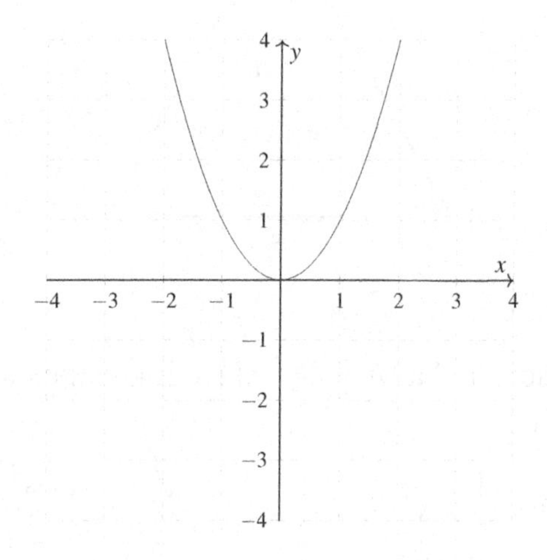

If $h(x)$ is shifted down by 2 units, the graph of $f(x)$ is obtained. Therefore, the value of m is -2 and the corresponding equation is $f(x) = h(x) - 2 \Rightarrow f(x) = x^2 - 2$.

14) To find the value of x, we solve the equation $5x - 2 = 3x + 8$. Subtracting $3x$ from both sides gives

$2x-2=8$. Adding 2 to both sides gives $2x=10$. Dividing by 2 gives $x=5$, which corresponds to option B.

15) The inverse function of $f(x)=\ln\sqrt{x^3}$ can be found by solving for x in terms of y. If we let $y=\ln\sqrt{x^3}$, then $e^y=\sqrt{x^3}$. Squaring both sides, we get $e^{2y}=x^3$, which means $x=e^{\frac{2y}{3}}$. Replacing y with x to find the inverse function, we get $f^{-1}(x)=e^{\frac{2x}{3}}$, which is option C.

16) First, solve $(x-3)^3=64$ by taking the cube root of both sides, giving $x-3=4$, so $x=7$. Then, substitute $x=7$ into $(x-7)(x-5)$, resulting in $(7-7)(7-5)$, which simplifies to $0\times 2=0$.

17) The function $y=3\left(\frac{1}{2}\right)^x$ is an exponential decay function. This is because the base of the exponent, $\frac{1}{2}$, is between 0 and 1. For such functions, as x increases, y decreases towards 0, but never actually reaches 0. Option A shows this behavior accurately: as x increases, y approaches 0. The other graphs do not represent this decay pattern. Options B and E shows exponential growth, the option C is a parabola, and the option D represents a linear function.

18) To find the area of an equilateral triangle with side length s, split the triangle by drawing an altitude to create two $30^\circ-60^\circ-90^\circ$ right triangles.

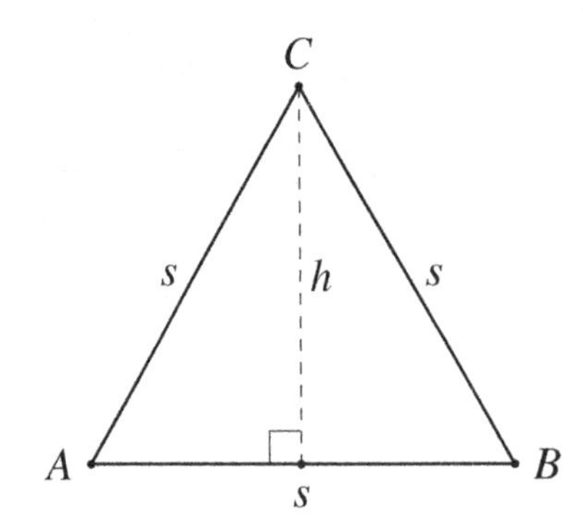

The altitude (h) is the longer leg, calculated as $h=\frac{s\sqrt{3}}{2}$ since it is opposite the 60° angle. The area A is given by:

$$A=\frac{1}{2}\times s\times h=\frac{1}{2}\times s\times\frac{s\sqrt{3}}{2}=\frac{\sqrt{3}s^2}{4}.$$

Thus, the correct formula for the area of the triangle is $\frac{\sqrt{3}s^2}{4}$, matching option C.

19) For the inequality $|x+5|\leq 7$, there are two cases to consider: $x+5\leq 7$ and $x+5\geq -7$. Solving the first gives $x\leq 2$, and solving the second gives $x\geq -12$. Combined, these solutions give $-12\leq x\leq 2$, which is option C.

20) To express 0.0048 in scientific notation, we move the decimal point three places to the right, making it 4.8. Because we moved the decimal to the right, the exponent on 10 is negative. Thus, 0.0048 is 4.8×10^{-3}, which corresponds to option C.

21) Start by expanding both sides: $\frac{5}{6}x - \frac{15}{6} = x - 10$. Then, simplify the equation: $\frac{5}{6}x - \frac{15}{6} = x - 10$. Combine like terms: $\frac{5}{6}x - x = -10 + \frac{15}{6}$. This simplifies to $-\frac{1}{6}x = -\frac{60}{6} + \frac{15}{6}$, which further simplifies to $-\frac{1}{6}x = -\frac{45}{6}$. Multiplying both sides by -6 gives $x = 45$, which is option E.

22) Factorize the polynomial:

$$9x^2 - 3x - 12 = 9x^2 + 9x - 12x - 12 = 9x(x+1) - 12(x+1) = (9x - 12)(x+1) = 3(3x-4)(x+1).$$

Thus, $3x - 4$ is indeed a factor, confirming option C.

23) The expression can be simplified as:

$$\left(xy^{-2}\right)^3 \left(\frac{y}{x}\right)^{-9} = \left(x^3y^{-6}\right)\left(\frac{x^9}{y^9}\right) = x^{3+9}y^{-6-9} = x^{12}y^{-15}.$$

So, $n - m = 12 - (-15) = 27$. Thus, the value of $n - m$ is 27.

24) The equation can be rewritten as $c - e = ac$. Divide both sides by c: $1 - \frac{e}{c} = a$. Since $c < 0$ and $e > 0$, the value of $-\frac{e}{c}$ is positive. Therefore, 1 plus a positive number is greater than 1. So, a must be greater than 1: $a > 1$.

25) To convert $6x - 4y < 3y + 42$ to y on one side, add $4y$ to both sides to get $6x < 7y + 42$. Then, subtract 42 from both sides to get $6x - 42 < 7y$. Divide everything by 7 to isolate y: $y > \frac{6}{7}x - 6$.

26) The function $g(x)$ is undefined when the denominator equals zero: $(x-4)^2 - 9 = 0$. This can be rewritten as $(x-4)^2 = 9$. Squaring both sides we have $x - 4 = \pm 3$. Thus, $x = 7$ and $x = 1$ are the values that make $g(x)$ undefined.

27) For the equation $3m^2 + 18m + 27 = 0$, divide all terms by 3 to simplify: $m^2 + 6m + 9 = 0$. This is a perfect square trinomial $(m+3)^2 = 0$, so $m = -3$. Since both solutions are -3, their sum is $-3 + (-3) = -6$, which corresponds to option C.

28) Substitute $t = 8$ into the equation to find g: $8 = \frac{(80-2g)}{0.5}$. Multiplying both sides by 0.5 gives $4 = 80 - 2g$. Rearranging the equation, we get $2g = 80 - 4$ which simplifies to $2g = 76$. Dividing by 2, we find $g = 38$. So, the gardener completed 38 bushes that week, which corresponds to option B.

29) The function $y = bx^2 + d$ with b positive and d negative will be a parabola that opens upwards (because b is positive) and has its vertex below the x-axis (because d is negative). Among the given graphs, only the graph in option A, matches this description. It opens upwards and has its vertex below the x-axis, indicating that it is the graph of a function where $b > 0$ and $d < 0$.

30) To isolate v in the equation $h = -t^2 + vt + 2s$, we can rearrange it to $vt = h + t^2 - 2s$. Then, divide both sides by t to get $v = \frac{h+t^2-2s}{t}$. To match the format of the options, this can be rewritten as $v = \frac{h-2s}{t} + t$, making option A the correct answer.

31) The transformation $g(x) = f(x) + 3$ includes a vertical shift upwards by 3 units. This affects the y-intercept but does not change the vertex's horizontal position or reflect the graph. Therefore, the y-intercept of $g(x)$ is 3 units above the y-intercept of $f(x)$, making option D correct.

32) Triple the quantity $-2x^2 - 3x + 6$ is $-6x^2 - 9x + 18$. Subtracting $7x^2 - 4$ from this gives $(-6x^2 - 9x + 18) - (7x^2 - 4)$. Simplifying, we get $-6x^2 - 9x + 18 - 7x^2 + 4$, which results in $-13x^2 - 9x + 22$, corresponding to option B.

33) To solve $(x-3)^2 + 2 > 4x + 6$, simplify the inequality: $(x-3)^2 > 4x + 4$. Expanding the square, we get $x^2 - 6x + 9 > 4x + 4$. Rearranging, $x^2 - 10x + 5 > 0$. We can test the provided options to see which one satisfies the inequality. Plugging $x = 0$ into the inequality shows that it holds, making option A correct. The other options did not satisfy the inequality when plugged into the equation.

34) To simplify $(4x-6)^2$, expand the square: $(4x-6)(4x-6)$. Using FOIL (First, Outer, Inner, Last), we get $16x^2 - 24x - 24x + 36$, which simplifies to $16x^2 - 48x + 36$, matching option B.

35) To determine the best model, observe the pattern of increase. The initial value (when $x = 1$) is 150, which matches the coefficient in options C and D. However, since the number of items sold is increasing, the base of the exponent must be greater than 1, eliminating option D. Option C has the correct initial value and a reasonable growth factor of 1.10.

36) To find the zeroes of the function $h(x) = x^2 - 6x - 7$, we need to factor the quadratic. In fact, we need two numbers whose product is -7 and sum is -6. These numbers are -7 and $+1$. Thus, we can factor $h(x)$ as $(x-7)(x+1)$. Setting each factor equal to zero gives $x - 7 = 0$ (so $x = 7$) and $x + 1 = 0$ (so $x = -1$). Therefore, the zeroes are 7 and -1, corresponding to option B.

37) To evaluate the y-intercept, we determine the coordinates of the intercept of the line graph with the y-axis. According to the graph, the line intersects the y-axis at point $(0, 3)$. So, choice B is equivalent to the y-intercept.

38) Remember that an exponential function has one horizontal asymptote. The horizontal asymptote of the exponential function is the horizontal line (the choices R or T can be the suitable answer) that becomes close to it for large negative or positive x. Clearly, the above exponential function graph gets close to the line T for the large negative x.

39) To find the equivalent function, expand the given function $g(x) = 3(3-4x)^2 - 9$. First, expand $(3-4x)^2$ to get $9 - 24x + 16x^2$. Then, distribute the 3 to get $27 - 72x + 48x^2$. Subtract 9 to get $18 - 72x + 48x^2$, which simplifies to $48x^2 - 72x + 18$. Therefore, the equivalent function is $g(x) = 48x^2 - 72x + 18$.

40) The x-intercept is the point where the graph crosses the x-axis, which occurs at $y = 0$. According to the graph, this happens at $x = 1$. Therefore, the statement that the x-intercept is 1 is supported by the graph.

41) The function $j(x) = h(x-2) - 1$ represents a transformation of the quadratic function h. The transformation $x - 2$ inside the function h indicates a horizontal shift of h units to the right, and the constant term -1 outside the function h indicates a vertical shift of 1 unit downward. Therefore, the vertex of the transformed function j is shifted 2 units to the right and 1 unit downward from the vertex of h. Since the vertex of h is $(-2, -2)$, the vertex of j is $(0, -3)$. Among the given options, graph A is the only one that has a vertex at $(0, -3)$, indicating that it best represents $j(x)$.

42) In the equilateral triangle if a is the length of one side of the triangle, then the perimeter of the triangle is $3a$. Then $3a = 36 \Rightarrow a = 12$ and the radius of the circle is $r = 12$. Then, the circumference of the circle is:

$$2\pi r = 2(3.14)(12) = 75.36.$$

43) Apply the formula $(a-b)^2 = a^2 - 2ab + b^2$:

$$(3x-4y)^2 = (3x)^2 - 2\times 3x \times 4y + (4y)^2 = 9x^2 - 24xy + 16y^2.$$

So the answer is C.

44) Take the natural logarithm of both sides: $\ln(e^{2x}) = \ln(25)$. This simplifies to $2x = \ln(25)$, so $x = \frac{\ln(25)}{2}$. Thus, D is the correct answer.

45) Multiply numerator and denominator by the complex conjugate of the denominator: $\frac{5-2i}{3i} \times \frac{-3i}{-3i} = \frac{-15i-6}{9} = \frac{-6}{9} + \frac{-15}{9}i = -\frac{2}{3} - \frac{5}{3}i$. Thus, the correct answer is D.

46) Use the formula of a quadratic function $y = a(x-h)^2 + k$, where the point (h,k) is the vertex and a is constant. According to the graph, the ordered pair $(1,2)$ is the vertex. Then, $y = a(x-1)^2 + 2$. To find the constant a, substitute a point like $(0,1)$ on the graph in the obtained equation. So,

$$1 = a(0-1)^2 + 2 \Rightarrow 1 = a + 2 \Rightarrow a = -1.$$

Now, plug $a = -1$ into the equation $y = a(x-1)^2 + 2$, and simplify. Therefore,

$$y = (-1)(x-1)^2 + 2 \Rightarrow y = -x^2 + 2x + 1.$$

47) To find the solution, first expand and simplify the equation: $2n - 2 = 3n + 6 - 10$. Simplifying the right side gives $2n - 2 = 3n - 4$. Then, subtract $3n$ from both sides to get $-n - 2 = -4$. Adding 2 to both sides gives $-n = -2$, and finally, dividing by -1 gives $n = 2$.

48) To solve $2x^2 - 11x + 8 = -3x + 18$, first bring all terms to one side to get $2x^2 - 8x - 10 = 0$. Factoring this quadratic equation, we have: $2x^2 - 8x - 10 = 2(x^2 - 4x - 5) = 2(x-5)(x+1) = 0$. Solving for x, we find the roots (solutions) are $x = 5$ and $x = -1$. Since $a > b$, we assign $a = 5$ and $b = -1$. Therefore, the ratio $\frac{a}{b} = \frac{5}{-1} = -5$.

49) The function $f(x) = 4(1.25)^x$ is an exponential function where the base, 1.25, is greater than 1. This means as x increases, $f(x)$ increases, making the function an increasing function (option E). The other options are incorrect because:

A) The function does not pass through $(-1,2)$,

B) The asymptote equation is $y=0$, not $x=0$,

C) The function does not have an x-intercept since it never touches the x-axis,

D) The function does not have an axis of symmetry.

50) Solve $x^2-16=9$ to find $x^2=25$, so $x=5$ or $x=-5$. Substituting $x=5$ into $(x-6)(x-3)$ gives $(5-6)(5-3)=(-1)(2)=-2$. Substituting $x=-5$ gives $(-5-6)(-5-3)=(-11)(-8)=88$. Since -2 is an option and 88 is not, the correct answer is -2, which is option E.

51) A strong positive association in a scatter plot is indicated when the points are closely clustered around a line that slopes upwards from left to right. This means as x increases, y also increases. In the given options, option A shows this pattern clearly with the points aligning closely along an upward sloping line, demonstrating a strong positive correlation between x and y.

52) Substitute $x=5$ into the equation: $y+2=3(5-4)$. This simplifies to $y+2=3$, so $y=1$. Thus, the y-coordinate of B is 1.

53) To find the y-coordinate where the graph crosses the y-axis, substitute $x=0$ into the equation: $y=-2(0)^2+8(0)+4=4$. Thus, the graph crosses the y-axis at the point $(0,4)$.

54) The range of a function is the set of all possible output values (y-values) it can produce. From the graph, the lowest y-value is -2 (at the point with a filled circle, indicating that -2 is included in the range), and the highest y-value the function reaches is 2. Since the graph includes a filled circle at $y=2$, 2 is also included in the range. Therefore, the range of the function is $\{y|-2\leq y\leq 2\}$.

55) Since Eleanor doubles the number of people she invites each year, the function representing the number of people invited grows exponentially. In the first year, she invites 50 people, in the second year 100, in the third 200, and so on, following the pattern $f(n)=50\cdot 2^{n-1}$ where n is the number of year. This pattern of doubling is a hallmark of exponential growth, making the function an increasing exponential function.

56) The shaded area represents the region where both inequalities are satisfied. From the graph, we observe that the solution to the system of inequalities is the intersection of the regions below the line $y=2x+1$, and above and on the line $y=-x$. Quadrant III contains no solutions.

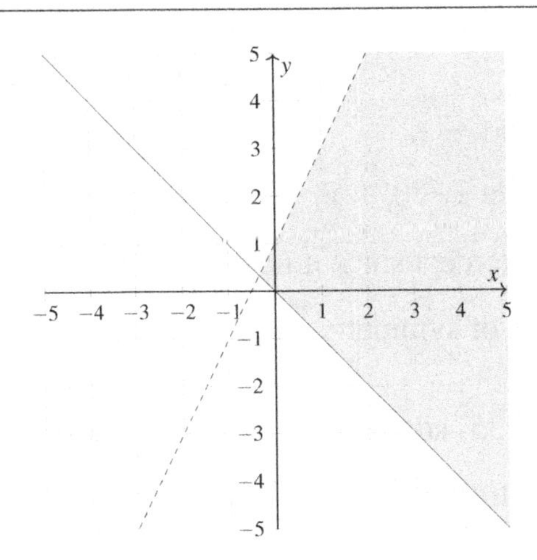

57) In the equation $y = 5x + 7$, the term $5x$ represents a quantity that increases by 5 times whatever x is, and the $+7$ represents a constant addition to that quantity. Option B aligns with this as for each group (x), there are 5 people, and then 7 more people are added in total. Option A is incorrect because the number of cookies would be $12x + 7$. Option C is incorrect because the cost would be $5x - 7$. Option D is incorrect because the conversion from feet to inches should involve multiplication by 12. Hence, B is the correct answer.

58) The expression $64x^2 - 16$ is a difference of squares and can be factored as $(8x)^2 - 4^2 = (8x + 4)(8x - 4)$. This matches option A.

59) Substitute $2x$ into $g(x)$: $g(2x) = 4(2x) + 3(2x + 2) + 5 = 8x + 6x + 6 + 5 = 14x + 11$. Therefore, the correct expression is $14x + 11$.

60) The function f has an x-intercept of 1 and a y-intercept of -4. This indicates that the line should intersect the x-axis at $x = 1$ and the y-axis at $y = -4$. Option A in the graph displays a line that intersects the x-axis at $x = 1$ and the y-axis at $y = -4$, aligning with the given intercepts. Therefore, this graph accurately represents

Author's Final Note

I hope you enjoyed this book as much as I enjoyed writing it. I have tried to make it as easy to understand as possible. I have also tried to make it fun. I hope I have succeeded. If you have any suggestions for improvement, please let me know. I would love to hear from you.

The accuracy of examples and practice is very important to me. We have done our best. But I also expect that I have made some minor errors. Constant improvement is the name of the game. If you find any errors, please let me know. I will fix them in the next edition.

Your learning journey does not end here. I have written a series of books to help you learn math. Make sure you browse through them. I especially recommend workbooks and practice tests to help you prepare for your exams.

I also enjoy reading your reviews. If you have a moment, please leave a review on Amazon. It will help other students find this book.

If you have any questions or comments, please feel free to contact me at drNazari@effortlessmath.com.

And one last thing: Remember to use online resources for additional help. I recommend using the resources on `https://effortlessmath.com`. There are many great videos on YouTube.

Good luck with your studies!

Dr. Abolfazl Nazari

Made in the USA
Las Vegas, NV
10 July 2024